Qanats and Historic Structures in Persia

Qanats and Historic Structures in Persia presents the early history of water science and includes the advanced knowledge held by Persians regarding the hydrologic cycle in general and groundwater flow in particular. It explains how the Persians understood the sources of rivers, streams, springs, and groundwater, at least seven centuries before it was known to western scholars, and how their use of underground water tunnels allowed them to transform deserts into centers of civilization and food production for thousands of years. It also presents an overview of ancient canals, weir bridges, dams, water storage structures, and water dividers constructed to supply water for irrigation and domestic needs.

- Presents numerous examples of how qanats are used throughout the world, including the Middle East, Africa, Asia, Europe, and South America.
- Includes descriptions and photographs of historic structures, some of which are still operational after hundreds of years.
- Written in an accessible and informative way, the book contains neither equations nor rigorous technical material.
- Examines the renowned scholars of the late ninth through twelfth centuries, namely the Persian Golden Era.

Hormoz Pazwash, PhD, earned his BSCE with the highest honor among the entire graduating class of 1963 from Tehran University. He continued his graduate studies under the supervision of the late Dr. Ven Te Chow at the University of Illinois, in Urbana-Champaign, earning an MS and PhD in civil engineering. In 1970, he joined the Faculty of Engineering at Tehran University, and in the next seven years, held the positions of assistant professor, associate professor, and chairman of the Department of Civil Engineering. His other academic appointments have included visiting professorships at Akron in Ohio (1978–79) and an associate professorship at Northeastern University in Boston (1982–85). He has also served as an adjunct professor at Stevens Institute of Technology in Hoboken, New Jersey.

Dr. Pazwash has received various academic awards, including a fellowship at Tehran University and a Fulbright Scholarship at the University of California, Berkeley. He has been engaged in engineering practice since 1986, and he retired in 2021.

Dr. Pazwash is the author of some 60 papers and five books, including *Urban Storm Water Management, 2nd Edition* (2016), and also a chapter in the *Encyclopedia of Environmental Management* (2014). He is a lifetime member and Fellow of the ASCE (American Society of Civil Engineers) and a Diplomat of AAWRE (Academy of American of Water Resources Engineers), D.WRE.

Qanats and Historic Structures in Persia
Potential Modern Applications

Hormoz Pazwash

CRC Press
Taylor & Francis Group
Boca Raton London New York

CRC Press is an imprint of the
Taylor & Francis Group, an **informa** business

Designed cover image: Shutterstock

First edition published 2025
by CRC Press
6000 Broken Sound Parkway NW, Suite 300, Boca Raton, FL 33487-2742

and by CRC Press
4 Park Square, Milton Park, Abingdon, Oxon, OX14 4RN

CRC Press is an imprint of Taylor & Francis Group, LLC

© 2025 Hormoz Pazwash

Library of Congress Cataloging-in-Publication Data
Title: Qanats and historic structures in Persia : potential modern applications / Hormoz Pazwash.
Description: Boca Raton, FL : CRC Press, 2025. | Includes bibliographical references and index. |
Identifiers: LCCN 2024006554 | ISBN 9781032659923 (hardback) |
ISBN 9781032659947 (paperback) | ISBN 9781032659930 (ebook)
Subjects: LCSH: Qanats–Iran. | Hydraulic structures–Iran. |
Karajī, Muḥammad ibn al-Ḥusayn, -approximately 1016. Kitāb Inbāṭ al-miyāh al-khafīyah. |
Hydrology–Early works to 1800.
Classification: LCC TD404.7 .P39 2025 | DDC 627.0955–dc23/eng/20240415
LC record available at https://lccn.loc.gov/2024006554

ISBN: 978-1-032-65992-3 (hbk)
ISBN: 978-1-032-65994-7 (pbk)
ISBN: 978-1-032-65993-0 (ebk)

DOI: 10.1201/9781032659930

Typeset in Times
by codeMantra

Contents

Preface

Most people think that history of water technology refers only to experimental and mathematical knowledge as we know it today. They only think of "modern science" as that which developed after the seventeenth century. This gives a very misleading idea of the whole evolution of scientific and technological achievements by mankind. It is as we know a man only in his maturity ignoring that such maturity was the fruit of long years of childhood and adolescence.

As George Sarton, one of the most prominent science historians, states, "It is unfortunately true that many scientists lack a cultural background; and because of this, they do not like to look backward. It is a vicious circle. Why should they look that way if there is nothing for them to see? Their history of science does not even go as far as the seventeenth century. They are prone to believe that almost everything worthwhile was done in the nineteenth and twentieth centuries. Now in this, they are most certainly wrong."

Old civilizations had made significant advancements in water resources engineering in general and groundwater hydrology in particular. While the Chinese promoted water well technology, Persians articulated in the construction of qanat (qanat in Arabic). Through qanat construction, they brought groundwater from hillside aquifers to playa skirts in arid and semi-arid parts of Persian lands and built cities and centers of civilization, a civilization that is referred to as the "qanat civilization." In fact, qanat is one of the most technologically advanced, the most sophisticated, yet the most difficult works of ancient man. Even with today's state-of-the-art technology, construction of a qanat is a difficult and precise engineering task. For this reason, Tolman, who is known as the founder of groundwater hydrology in the United States, refers to qanat as the most extraordinary work of ancient man. In fact, qanat is also the most wondrous work of ancient civilization. It is far more wondrous than all wonders of the world, and undoubtedly, the most beneficial to human welfare and mankind than them all.

The most astonishing discoveries of the works of ancient civilization were achieved partly through efforts of interested researchers, but more so because of the need for survival. It is to be noted that all Persian writings prior to the seventh century AD were destroyed during the Arabs invasion. Upon conquering Ctesiphon (called Tysfun in Iran), the capital city of the Sasanids (Sasanian) Empire, as will be noted in this book, Arabs invaded the main library and destroyed all books and manuscripts. For nearly 250 years, they forced people to convert to Islam and speak in Arabic. Thus, Arabic became the sole literary language in Iran, and up to the early eleventh century, the time of Ferdousi Tousi (spelled as Ferdowsi in English literature), the great Persian epic poet who revived Farsi (or Parsi, the old Persian language), all works of Persian scholars, whether in medicine, science, mathematics, or engineering, were written in Arabic. Even the first book on Arabic grammar was written by a Persian named Sibvayh. (reads Sibwayh in English writings) What is wrongly known as the Islamic Golden Era is nothing but the works of Persians and is totally unrelated to that religion.

Recent discoveries of works of Persian scholars brighten our knowledge of science history. An excellent example is a text written by Mohammad Karaji, a Persian mathematician and engineer of the tenth and early eleventh centuries. A copy of one of his books titled *Extraction of Hidden Water* written in Arabic (the common language in Persia at the time) was discovered at Heiderabad Kan in the 1960s and was translated into Farsi (the Persian language) by Husayn Khadiv Jam in Tir mah 1345 (July 1966). The book covers subjects such as the occurrence of groundwater and surveying and leveling instruments (some of which he personally invented) for carrying out the construction of ghanats. Excerpts of Karaji's book which are presented in a chapter of this book reveal that Persians had a genuine knowledge of the origin of springs and rivers and the occurrence and movement of groundwater over a thousand years ago. In fact the state of knowledge of the science of hydrology and in particular groundwater flow as presented in that millennial book surpasses those of the Western scientists of the seventeenth and even the eighteenth century. The present book includes five chapters. The first chapter presents the early ideas about the elements of terrestrial matters, earth geometry and the origin of springs and rivers. xcerpts from a millennium book titled *"Extraction of Hidden Waters "*by Karaji on the origin of rivers and springs which are presented in this chapter reveal that Persians had genuine knowledge of the origin of springs and rivers hundreds of years before the Western scholars. This chapter also covers the Golden Eras of the Achaemenids and Sasanids (misspelled Sassanids in many English writings, Sasanian, in Persian) Empires and presents a brief presentation of Persian contributions in various fields of medicine, science, technology and engineering during the ninth through twelfth centuries. It is shown that what is wrongly known as the Islamic Golden Era was in fact the Persian Golden Era.

The second chapter includes excerpts from Karaji's book on the earth geometry, force of gravity, the origin of springs and rivers, and the occurrence and movement of groundwater in Persia. A review of his book titled *"Extraction of Hidden Waters"* which is the oldest book of hydrologic cycle and groundwater technology reveal that Persians knowledge of the science of hydrology, and in particular, groundwater flow preceded those of western scientists of the seventeenth and even the eighteenth centuries.

Chapter 3 discusses the origin of ghanats (qanats in Arabic, Kariz in Persian) and its spread in the world. It is shown that because of ghanats (underground water tunnels), Persians could turn deserts to fertile lands and centers of civilization over three thousand years ago. It is noted that qanat was the most ingenious and extraordinary work of ancient man. In fact, ghnat is the most wondrous of wonders of the ancient civilizations. It also includes a survey of ghanats which were built in Middle East, Far East, Africa, Europe and as far as South America.

The next chapter titled "Ghanat Construction" covers the search for a suitable location for qanat construction. This chapter includes descriptions presented in Karaji's book on the search for groundwater and procedures and instruments, some of which was invented by himself, to carry out qanat construction. Also included in this chapter are the procedures for the maintenance and repair of qanats and, their safe guarding and water rights.

Chapter 5, titled "Persians' Ancient Hydraulic Structures," presents an overview of ancient canals, weir bridges, dams, water storage structures (cisterns, ab-anbars, in Farsi), and water dividers (ab-pakhshans in Farsi) that were constructed by Persians to supply water for irrigation and domestic needs. Also included in this chapter are water measuring devices for allocating irrigation water.

About the Author

 **Dr. Hormoz Pazwash, PhD,** earned his BS, CE with the highest honor among the entire graduating class of 1963 from Tehran University. He continued his graduate studies under the supervision of the late Dr. Ven Te Chow (a well-known authority in the areas of Hydrology and Hydraulics) at the University of Illinois, in Urbana-Champaign, earning an MS and PhD in civil engineering. In 1970, he joined the Faculty of Engineering at Tehran University, and in the next seven years, held the positions of Assistant Professor, Associate Professor, and ultimately chairmanship of the Department of Civil Engineering. His other academic appointments have included visiting professorships at the University of Akron in Ohio (1978–79) and associate professorship at Northeastern University in Boston (1982–85). He has also served as an adjunct professor at Stevens Institute of Technology in Hoboken, New Jersey.

Dr. Pazwash has received various academic awards, including a fellowship at Tehran University and a Fulbright Scholarship at the University of California, Berkeley. For 36 years, from 1985 until his retirement in 2021, he has been engaged in engineering practice in the fields of hydraulic and hydrologic engineering and storm water management.

Dr. Pazwash is the author of some 55 papers and six books, including *Urban Storm Water Management,* 1st Edition in 2011 and 2nd Edition in 2016, and a chapter of the *Encyclopedia of Environmental Management*, published in September 2014. He is a lifetime member and a Fellow of the ASCE (American Society of Civil Engineers) and a Water Resources Engineer Diplomat of the AAWRE (American Academy of Water Resources Engineers).

1 Early History of Water Science and Persian Golden Eras

1.1 CONCEPTS REGARDING ELEMENTS OF MATTER

Man in antiquity thought that terrestrial matters were made of one or a combination of four elements: fire, air, water, and earth or a combination thereof. Thales of Miletus (580 BC) thought that the source of all things was water; Heraclitus (502 BC) thought it was fire; and Anaximenes (died ca 526) and Euripides (450 BC) believed that it was all four elements: fire, air, earth, and water. Empedocles affirmed that by composition of these four elements in different proportions all innumerable different substances known to man can be formed. Perhaps Pythagoras (530 BC), a predecessor of Empedocles, was the first to give up the idea of one single element as the basis of matter; he held the idea that matter is composed of earth, water, air, and fire. However, he thought that matters are derived from the combination of pairs of four governing qualities, hot and cold, wet and dry. Water for instance was cold and wet, while fire was hot and dry.

The idea of four elements apparently was derived from a misinterpretation of the natural action of fire. It was thought that a substance, when burned, resolves into its elements; the combustible matter is complex, but the small ash left by burning is simple. For instance, when burning green wood, the fire is seen by its own light, the smoke disappears in the air, water boils off from the wood ends, and the ash left is clearly the nature of the earth.

The idea of four elements was followed by later philosophers. Aristotle (384–322 BC), for example, pursued Empedocles but further defined the elements. He stated that fire has the highest and earth the lowest place among the four elements, with air being nearer to fire and water to earth. He asserted further that these elements originate from one another and each of them exists potentially in each. Centuries later, still many scientists and physicians accepted the four-element theory; some even tried to relate it to their work. Galen, a Greek physician of the second century, based much of his renowned works in medicine on the premise of four elements and particularly the characteristics attributed to them.

The concept of four-element theory was accepted not only by Greeks but also by the Muslim philosophers of the ninth and tenth centuries whose writings were primarily in Arabic, although most of these philosophers were non-Arab and in particular Persian. This concept was also the basis of much advancement in medical science in China. In the West, the concept of four elements prevailed until the publication of the *Skeptical Chemist* by Boyle[1] in 1661. Even during the same century, Johann Baptista Van Helmont, a Flemish chemist, conducted an experiment with a willow tree and concluded that water is the chief element of all matter. Over five years,

DOI: 10.1201/9781032659930-1

Helmont watered the tree but provided no other substance to the tree other than the soil it was initially planted in. But observing that the tree increased its weight and size, he attributed the gain to be solely due to water.

1.2 EARLY KNOWLEDGE OF EARTH GEOMETRY

1.2.1 WESTERN THOUGHTS

To put the Persian knowledge in perspective, an overview of western ideas is first presented in this section.

Many ancient cultures, including the Greeks, believed that the earth was a flat disk until the Hellenistic period (323–31 BC).[2] The idea of spherical earth was introduced by Pythagoras (sixth century BC); however, his idea lacked any justification and Greeks retained the flat earth idea until Aristotle (384–322 BC) provided empirical evidence of the spherical shape of the earth. The paradigm of spherical earth was gradually adopted throughout the Late Antiquity and Middle Ages. A practical demonstration of the earth's sphericity was provided by Ferdinand Magellan (a Portuguese explorer) and Juan Sebastian Elcano (a Spanish explorer) in expedition circumnavigation (AD 1519–1522).

The idea of flat earth cosmography prevailed in India until the Gupta period (early centuries AD) and in China until the seventeenth century. In early Mesopotamian mythology, the world was conceived to be a flat disk floating in the ocean and surrounded by a spherical sky. Another speculation on the shape of the earth is a seven-layered ziggurat or cosmetic mountain (seven climes) held in the Avesta and ancient Persian writings.

Though the earliest writing on the spherical earth was speculated by ancient Greeks, the spread of the idea of a spherical Earth in Europe was very gradual due to the pace of Christianization of Europe. For example, the first evidence of the spherical shape of the earth in Scandinavia is a twelfth-century Old Icelandic translation of "Elucidarius" (the obscurity of various things). However, the spherical shape of the earth was known to seafarers many centuries earlier. (See also Asfazari's description of ocean shape in the next section.)

1.2.2 PERSIAN KNOWLEDGE OF EARTH GEOMETRY

As will be indicated later in this chapter, Persians had two Golden Eras of science and engineering, from the fifth century BC to the early seventh century AD. The first era occurred during the Achaemenid (Hakhamaneshian) Empire from 550 to 330 BC. Centuries later, the Sasanid (Sasanian) Empire (AD 224–651) witnessed the peak of ancient Iranian civilization before the Arab's conquest from 633 to 635, and the enforced adoption of Islam religion. However, all Persian writings of their first and second Golden Eras were destroyed by the order of Caliph Umar during the early stages of the Arab's invasion of Persia in the seventh century.

The third Golden Era began when the Arabs' domination of Iran was practically ended near the end of the ninth century and continued for nearly three centuries through the mid-twelfth century AD. This era, which is improperly referred to as

Islamic Golden Era is indeed the third Persian Golden Era. As will be indicated later, Persians made significant contributions to science, mathematics, medicine, and engineering in this era. Alas, many of the Persian writings of their third Golden Era were lost during attacks by Mahmud of Ghazna and Turco-Mongols. Only a fraction of the books composed by Persian mathematicians, physicians, and engineers of the mid-ninth through twelfth centuries have survived. One of these books is a late tenth to early eleventh-century manuscript on groundwater hydrology which forms a prime subject matter of two chapters of this book. The manuscript, written by Mohammad Karaji, a Persian mathematician and engineer who lived from AD 953 to 1029, is undoubtedly the oldest treatise on hydrology (Nadji and Voight 1972; Pazwash and Mavrigian 1980, 1981). The book presents a genuine description of the earth, the concept of the hydrologic cycle, the source of springs and rivers, the movement of groundwater, and the means of its extraction. Excerpts of this manuscript, entitled *Extraction of Hidden Waters*, are presented in the next chapter. Only Karaji's writing relative to the subject matter of this chapter is briefly presented herein.

Abu Bakar Mohammad Karaji, a brilliant mathematician and engineer of the mid-tenth to early eleventh centuries AD, in his famous book *Extraction of Hidden Waters* writes:

> Earth with all its mountains, plains, hills and valleys; has a spherical geometry. It is a tiny element of the universe and moves around its center. The God placed it at the center of the universe and it turns around this center forever (Karaji, p.3).

Also, until the seventeenth century, it was believed that the earth is stagnant and the sun rotates around it. Galileo Galilei (AD 1564–1642) was the first Roman scholar who said that the earth moves around the sun. For this Copernican heliocentric idea, he was sentenced to house arrest by the church for the rest of his life.

Abu Hatim Muzaffar ibn Esmail Asfazari is another Persian scientist and astronomer of mid-eleventh, early twelfth century AD (mid-fifth, early sixth century AH[3]) who knew about the earth's geometry. Asfazari lived about the same time as Omar Khayyam and nearly a century after Karaji. One of his books, written in Persian and titled *Ressaleh Assar Alavi* (Natural Sciences Treatise), was edited by Mohammad Taghi Modarres Razavi in 2536 Shahanshahi (1356 SH,[4] AD 1977). Included in the book are chapters on the origin of rain and snow, the formation of dew and winds, rainbows, and geyser.

On sphericity of seas and oceans, Asfazari (1977) writes:

> Vast bodies of water are not flat; rather they are curved like the surface of a sphere.[5] For this reason, the sailors approaching the cost, first see the top of trees along the coast; and as they come closer, they see the trees completely. Likewise, people standing on the coast first see the top of a ship approaching the coast; and as the ship gets closer, they see the entire ship. And, as the ship is moving away from them, they first see the whole ship and then they can see only its top. *(Asfazari, pp. 42–43)*

The earth's circumference was estimated by Claudius Ptolemy (AD 100–170) who lived in Alexandria, the center of scholarship in the second century. His estimate was 89.7 km/degree (which is equivalent to 5,139 km radius). A more accurate estimate was made by Ahmad Farghani, a Persian astronomer of the ninth century. Abu Rayhan Biruni (AD 973–1048), a Persian scholar, astronomer, geographer, mathematician, and historian who is called the founder of geometry, used a new method involving trigonometric calculations to accurately compute the earth's circumference (Aber 2003). His estimate of 6,339.9 km for earth's radius was only 16.8 km (0.26%) less than the modern value of 6,356.7 km.

Until the Renaissance, western scholars generally believed that the earth is stagnant and the sun circles around it. Abu Rayhan Biruni, jointly with his colleague, Abu Sahl (AD 940–1000, a Persian mathematician, physicist, and astronomer from Amol in Tabaristan) wrote a treatise on this matter. He questioned whether the earth is stagnant or it is moving. Apparently, he did not believe that the earth is stagnant. Unfortunately, that writing has been lost. Also, as indicated earlier, Karaji writes: "The God created it (the earth) at the center of the universe and it turns around this center forever."

1.3   IDEAS ON THE ORIGIN OF RIVERS AND SPRINGS

1.3.1   IDEAS IN THE WEST

Ancient Greek philosophers, such as Thales (640–546 BC), Plato (427–347 BC), and Aristotle (348–322 BC), speculated the origin of rivers and springs. They were impressed with the large size of rivers compared with the amount of rain. They were also amazed by the caves and large springs in the Balkan Peninsula. However, the ideas and theories they presented for the origin of springs and rivers were generally faulty. Plato, for example, theorized large underground caverns (Tartarus) as the source of rivers and springs. Aristotle suggested the condensation of air into water in cold mountains to be the source of springs and mountain streams. Romans generally followed Greek ideas. Seneca (4 BC–AD 65), for instance, held essentially the same theory as Aristotle on the source of springs.

In the Middle Ages and early years of Renaissance, the western scientists such as Kepler, Kircher, and Magnus exhibited great interest on the subject of groundwater but expressed ideas that were essentially different from those of our basic knowledge today. For instance, Johannes Kepler (AD 1571–1630), a German astronomer and mathematician with a great appetite for imagination, offered this illusive idea "the earth drinks water from the sea and digests it like a beast; and as a result, springs originate."

The gradual change from purely philosophical concepts toward observational and measurement activities occurred during the Renaissance near the end of the fifteenth century AD.

Leonardo da Vinci (AD 1452–1519), the most outstanding figure of the Renaissance and genius of Rome, was the first westerner who presented a somewhat correct explanation for the origin of streams. He also made velocity measurements of streams using a weighted rod held suspended by an inflated animal bladder (MacCurdy 1939). Bernard Palissy (1510–1589), a French scientist and another important figure of this era, reached some modern ideas on the origin of rivers in the form of a rather lengthy dialogue between "theory" and "practice."

Pierre Perrault (AD 1608–1680), a French hydrologist, Edme Mariotte (1620–1684), a French physicist and priest, and Edmond Halley (1656–1742), an English astronomer and meteorologist, were the most important western figures of seventeenth and eighteenth centuries on the history of hydrology. These three men based their ideas on measurements and for the first time on a quantitative basis. Perrault measured the amount of rainfall on the River Seine basin and found that it was about six times the volume of water discharged by the river. Mariotte measured the infiltration of rainwater in the basement of a Paris observatory and noted that it varied with the amount of rainfall. Halley made crude tests of evaporation from the Mediterranean Sea and demonstrated that the evaporation was sufficient to account for all the water supplied to the streams and springs. In the same period, Antonio Vallisneri (in 1715), an Italian naturalist, explained the mechanism of artesian wells. However, the first correct explanation for such mechanism was given by brilliant Persian physicist, mathematician, astronomer, and natural scientist, Abu Rayhan Biruni[6] (973–1048), seven centuries earlier. Biruni's picture has been a motto of Russian and Afghan stamps. Figure 1.1 shows his picture on a Russian stamp.

1.3.2 Persian Knowledge of the Source of Rivers and Springs

On the source of springs and rivers, Karaji (p. 7) states,

> The origin of all waters inside the earth and the source of springs, rivers, streams and all water bodies that exist on the earth is the rain and snow. If the rain and snow cease to fall, water depletes and the earth eventually desolates.

These statements clearly evidence that Persians knew about the source of rivers and springs at least seven centuries before Pierre Perrault's finding on the River Seine in the seventeenth century.

On the origin of rivers, Asfazari states,

> As the accumulated snow on the mountain melts, it forms small streams; and as these streams are joined, larger streams are created; and when these large streams merge, a large river is formed. Rivers, thus created, are perennial and keep flowing through the summer. Since the source of these rivers is snow, their flow increases as the weather becomes warm. However, those rivers which originate from such water, dry up in summer (Asfazari, pp. 37–38).

FIGURE 1.1 A Russian 1973 stamp featuring Abu Rayhan Biruni.

In addition to these scientists, the origin of springs and rivers are described in a number of literary works. One such example is a verse from Golestan (Flower Garden),[7] a literary work of Saadi of Shiraz (AD 1210–1292), who is known world-wide as one of the world's greatest poets, writer of literature and philosopher.[8] This verse translated by the author also proves that Persians had a genuine knowledge of the source of rivers and streams centuries before Europeans.

> If your income is low, lower your expenses As a song changed by sailors says:
> If no rain on the mountain falls,
> the Diyala River that year fully dries.

Asfazari on geysers writes:

> When a large amount of water steam is formed underground and this vapor reaches a porous surface, it discharges like a wind. However, if the ground is tight and has no opening to let the vapor to escape, the vapor keeps building up and rising because of heat until it reaches the ground surface, it shakes the ground to create an opening and gushes out in the air. A whistling noise is heard before the ground is broken and the water blows up. Asfazari refers to this phenomenon which is geyser as "hot spring." And as the vapor loses its heat and cools down, its power diminishes and the geyser disappears. And when the warm vapor builds up again, this process repeats itself. This resembles wind blowing continuously from a hole in a mountain. (pp. 40–41, Chapter 2, Sec. 4)

1.4 SOURCE OF GROUNDWATER

1.4.1 INFILTRATION

Up to the mid-seventeenth century, the western scholars generally believed that infiltration was inadequate to supply water to springs. Marcus Vitruvius Pollio (80/70–15 BC), a Roman military engineer of Julius Caesar, may be an exception to this. He was the first western scholar who refers to qanats of Persia and its source being the rain and snow on the mountains. In his text "On Architecture" he notes that "as melting snows through the interstices of the earth and reach to the lowest spurs of the mountain from which the springs flow." However, the relation between infiltration and streams and springs was unknown to western scientists until the Renaissance.

In Persia, on the other hand, infiltration as an element of the hydrologic cycle was a known subject as early as the tenth century AD and undoubtedly centuries earlier. The construction of qanats, which is presented in a later chapter, indicates that the Persians were aware of infiltration as the origin of groundwater over three thousand years ago. On the subject of infiltration, Karaji in his *Extraction of Hidden Waters* states:

> Water from the snow melt enters the veins and cracks which exist inside the earth, flows downward and becomes the source of water at far away places. *(Karaji, p. 6)*

More on Karaji's knowledge is presented in the next chapter of this book.

On infiltration, Asfazari writes:

As melting snow from a mountain reaches a gravely and sandy area, the water percolates the soil and disappears. And, as the water travels underground, it accumulates until it finds an opening on the ground to escape. There it discharges as a spring. However, if the water can not find an escape route, it remains inside the earth as soil moisture. And, if along its route, there are any substances such as salt or sulfur inside the ground, the water mixes with that substance and carries its odor and taste to where it discharges as a spring. If the water is highly concentrated with salt and its flow gradually drops, salt builds up at the spring bed and hardens as a rock overtime. *(Asfazari, Chapter 2, Section 3, pp. 38–40)*

1.4.2 GROUNDWATER FLOW

In the nineteenth century AD, Henry Darcy (1803–1858), a French scientist who was inspired by the test results of Hagen and Poiseuille on laminar flow in pipes, studied the flow of water through sand. Based on this study, he for the first time expressed the mathematical law that governs the groundwater flow. This law, which is known as Darcy's Law, is expressed by an equation that indicates the flow velocity is proportional to the water table gradient. The proportionality factory, which is called permeability or hydraulic conductivity, depends on the soil type.

On groundwater flow, Karaji writes:

Ground waters, analogous to surface waters, are flowing as streams at places and are stagnant as a sea at others. Stagnant waters are mostly located beneath vast deserts and low lands and can be tapped at definite depths. *(Karaji, p. 15)*

Through experience, the Persians knew that constructing qanats (underground conduits which bring water from foothills to playa skirts[9]) at steep slopes creates high velocity and causes erosion of the conduit. Though they did not explicitly express the linear relation between the groundwater flow velocity and the grade, they knew that higher the grade, the greater the velocity. Therefore, the qanat conduits were dug along a very mild slope (commonly at a 0.10% grade) to avoid soil erosion. They also knew that at vast flat plains where the water table has no gradient, the groundwater becomes stagnant.

1.5 IRRIGATION CANALS

It is not known where the first irrigation system was constructed. There is evidence of the existence of dams and irrigation canals in Mesopotamia and Egypt five thousand years ago. In ancient Egypt, the construction of canals was a major endeavor of Pharaohs and their servants since King Scorpio's time, ca 3200 BC (Mays 1996). King Menes, also spelled Mena or Min (ca 3000 BC) is said to have constructed a diversion dam on the Nile to bring water to Memphis, his capital. Three centuries later, about 2700 BC, Sad-el-Kafara Dam was built twenty or so kilometers south of Cairo. This dam, apparently, had no spillway and failed during the first rainy season after its construction (Linsley, Kohler and Paulhus 1982, p. 3).

Irrigation was not a major concern in Rome because of the terrain and intermittent streams. The source of a typical water supply system of a Roman city was a spring or a dug well. The water was lifted from a well or spring or several fresh sources and was brought to the town by aqueduct. The early Roman bourgeois family typically had a large house with a hole in one roof to let rain in and get stored in a cistern in that room.

The ancient irrigation systems in Mesopotamia and Egypt Delta were of the basin type which were opened by cutting a gap in the embankment and closed by putting mud back into the gap. Laws in Mesopotamia required farmers to help maintain their basins, dams, and feeder canals in repair and new canals were dug (de Camp 1963). Some canals were in use for centuries before they were abandoned and new ones were built.

Hammurabi (eighteenth century BC), the sixth King of the Amorite dynasty of Old Babylon, has the oldest Code of the Middle East. Some of his codes were pertaining to slavery and women engaged in business and were primitive and brutal.[10] Relating to dams, his code reads (translated by L. W. King, 2005):

> If anyone be too lazy to keep his dam in proper condition and does not so keep it; if then the dam breaks and all the fields be flooded, then shall he whose dam the break occurred be sold for money and the money shall replace the grain which he has caused to be ruined.

The Greeks borrowed ideas from the Babylonians, Egyptians, and Phoenicians. The Sumerians and Akkadians (including Assyrians and Babylonians) dominated Mesopotamia from the beginning of written history (ca 3100 BC) to the fall of Babylon in 539 BC, when it was conquered by Cyrus the Great of Achaemenid Empire. Mesopotamia included the Tigris-Euphrates River system. The land between these rivers which was referred to as Miyan Rudan (between rivers) by Persians, included parts of present Syria, all of Iraq and southeastern Turkey. The said rivers unite at their extreme south reaches and empty to the Persian Gulf.

Mesopotamia fell to Alexander III of Macedon (336–323 BC), known as Eskender the Great in the West and Alexander Maghdoni in Iran; and after his death, it became part of the Greek Seleucid Empire. Around 150 BC, Mesopotamia came under the control of Persian Parthian Empire and became a battleground between the Romans and Parthians and part of it lost to Romans. In AD 226, the Persian Sasanid (Sasanian) Empire took control of the entire Mesopotamia and remained under their control until the Muslim invasion of Persia, which began in AD 632.

Mesopotamia is the site of the earliest developments of the Neolithic Revolution and has inspired some of the most important developments in human history, including irrigation and agriculture. The irrigation in this area was aided by melting snows from the high peaks of northern Zagros Mountains (in Khuzestan Province) and the Armenian Highlands and high water table. This area reached its peak prosperity and scientific advancement during Sasanian Empire who chose Ctesiphon[11] (called Tysfphon or Tyspfon in Persian, Madáin in present Iraq) as their capital city and constructed Taq e Kesra (Ayvan e Kesra, meaning Iwan Khosrow) in that city near the Tigris River.

1.6 GOLDEN ERAS OF SCIENCE AND TECHNOLOGY IN ANCIENT PERSIA

1.6.1 HAKHAMANESHIAN EMPIRE

The period of Achaemenids (Hakhamaneshian) Empire, which ruled for 220 years (550–330 BC), was the first Golden Era in Persian history. During this era, Persians built numerous qanats (qanats) in Persian territory which included Mesopotamia and extended to three continents. Through these qanats, Persians extracted fresh groundwater to provide water for irrigation and domestic needs thereby turning deserts into fertile lands and building cities and centers of civilization. They also built dams and canals to manage rivers for irrigation (see Chapter 5).

Cyrus the Great (known as Kurosh (or Kourosh) Bozorg in Iran) who lived from 600 to 529 BC, was the first Hakhamaneshian Emperor and the founder of Persia by uniting the two original tribes, the Medes and the Parthians. He conquered Median Empire, then Lydian Empire and lastly Babylonian Empire and founded the first Persian Empire to rule from the Balkans to North Africa and also central Asia. Wilhelm Friedrich Hegel, a German philosopher (1770–1831), names the Persians as the first historical people, the first empire and the only civilization in all history to connect over 40% of the global population of approximately 49.4 million around 480 BC. Cyrus the Great, who ruled from 559 to 529 BC, respected the customs and religions of the lands he conquered. He is generally more admired as a liberator than a conqueror. He wrote the first Charter of Human Rights[12] more than two millennia before the 1789 French Declaration of the Rights of Man and the Citizen. (see Ref. Cyrus Cylinder)

October 29 has been designated as the International Day of Cyrus the Great, King of Persia, who declared the first Charter of Human Rights. Cyrus the Great is known to have liberated the Jews from the Babylonia captivity in 539 BC to resettle and rebuild Jerusalem. On October 29, 539 BC, he himself entered the City of Babylon after his troops had peacefully entered the city and ordered his soldiers to

FIGURE 1.2 Kurosh's tomb in Pasargadae (the author in front, 1971).

respect people's belief. He is referred to by the Jewish Bible as Messiah. According to Xenophon, Cyrus created the first postal system in the world to help with infra Empire communications. Ecbatana (now Hamedan), Pasargadae,[13] Susa and Babylon were the centers of Cyrus's command. Cyrus is the Greek version of the old Persian Kurus' or Khurvas' meaning "sun-like," the noun Khúr denotes "sun" and vaś or vash is the suffix of likeness.[14] Kourosh was born in Perses, which roughly corresponds to the current Province of Fars.

Figure 1.2, taken in 1971, shows Kurosh's tomb in Pasargadae where he founded a new capital city. This structure was classified as a World Heritage Site in 2015. There is not a universal agreement on Kurosh's religion. The fire altars and the mausoleum at Pasargadae indicate his devotion to Zoroastrianism. Kurosh, according to Herodotus, was killed near the Aral Sea in July or August 529 BC during a campaign to protect the northern borders of his empire from invasion by Massagetae (ancient eastern Iranian nomads).

Hakhamaneshian, in general, and Kurosh, in particular, promoted the art, philosophy, music and architecture, and they have made a profound impact on both the Eastern and Western world. Persian philosophy, literature, and religion played a dominant role in world events of the next millennium. Despite the conquest of Persia by Islamic Arabs in the seventh century, Persians continued to have enormous influence in the Middle Ages during the so-called Islamic Golden Era, which as will be noted was indeed the last Persian Golden Era.

Dariush (Darius the Great), the third Hakhamaneshian King, as will be indicated in Chapter 3, built a qanat in Egypt and brought waters to the Oasis of Kharga (Butler 1933). He also built the Palace of Apadana (called Takht-e Jamshid in Iran), in Persepolis and Daryoon Canal and Salasel Castle in Shushtar (see Chapter 5). Apadana included a 100 m by 100 m center hall with tall large round stone columns. This building was destroyed by Alexander of Macedon (known as Alexander the Great in the West) upon defeating the last Achaemenid King in 330 BC. Only columns and pillars of pillars at the foyer of the palace which covered 625 m². have remained which present breathtaking views of remnants of this magnificent palace. About 5 km north of Persepolis there is another ancient monument called Naqsh-e Rostam. This monument contained several old graves when Dariush the Great (522–486 B.C) ordered his monumental tomb to be carved into the cliff. This tomb contained two inscriptions (DNa and DNb), which meant Dariush wanted to rule by justice:

> It is not my desire that a man should do harm, nor is my desire that he goes unpunished when he does harm.

1.6.2 SASANIAN: SECOND PERSIAN GOLDEN ERA

Sasanian Empire, which ruled for nearly 430 years from AD 224 to 651, followed suit of Achaemenid Empire. This empire, which formed the second Golden Era in Persia, also built many qanats, irrigation channels and canals, and weirs including a masterpiece of hydraulic engineering work in Shushtar in Khuzestan (see Chapter 5). on the old canal which was constructed by Daruish, the Great Achaemenid king, In fact, the Sasanian empire had two Golden Eras; the first one from AD 309 to 379 and the second during AD 498 to 622. The kings of this dynasty chose Ctesphon (Tisphon in Farsi) along the Tigris River as their capital. During King Khosrou I (spelled Khosrow in English writings, known as Anushirvan the Just in Iran), who ruled from AD 531–579, an archway to the city of Tisphon was built. The exact time of construction is not known with certainty, but appears to be around AD 540. This archway, which is named Taq Kasra (or Kesra) or Iyvan-e Kasra Taq-e Khosrow, Iyvan Madâen is open on the façade side, and is 35 m high, 43.5 m long and 25.5 m wide.

The arch of this structure is the largest single-span vault of unreinforced brick-work ever constructed in the world. The top of the arch is about 1 m thick while the walls at the base are up to 7 m thick. Taq-e Kasra, which is the only visible remaining structure Ctesiphon (Tisphon in Iran), is one of the wonders of the world. This structure has been studied and discussed by numerous people including French artist Eugene Flandin, who visited and studied the structure with Pascal Coste, a French architect, who remarked the Romans had nothing similar or of the type.

It is to be noted that because of Arabs invasion in the first half of the seventh century, no written documents of Golden Eras of Hakhamaneshian and Sasanian have survived. Not only did they destroy books but also demolished many historical structures and fire towers. They ignorantly considered fire towers as the houses of fire worshiping of Zoroastrians. Unknowing that Persians, unlike them, did not like darkness and had built fire towers to brighten the area at night to keep evils and ill actions at bay. The existing qanats and remnants of fire towers, irrigation canals, dams and weir structures illustrate this golden era of Persians before the Arab invasion in the seventh century.

Sasanian's territory stretched to the Mediterranean Sea and covered the present Egypt on the West and parts of China and India on the east. The Persian cultural influence during this dynasty extended beyond the empire territories reaching as far as Western Europe, Africa, China, and India. It played an important role in the formation of both European and Asian medieval arts.

Much of what later was known to westerners as Islamic culture in art, music architecture and other subject matter was transformed from the Sasanian through the Muslim world (Zarrinkoob 2005, 2017). Regrettably, few writing of the Achaemenid and Sasanid Golden Eras has survived.

Upon invading Persia during the AD 633 through 651 period, Arabs like today's ISIS, brutally killed Persians who were defending their homeland and/or refused to convert to Islam. They also destroyed all Persian books in the early stages of invasion. It has been said that Saad ibn Abi Waqqas, Caliph Umar's commander, upon capturing Ctesiphon (the capital city of Sasanian Empire)[15] at the Battle of Qadisiyyah in AD 636 and invading libraries, asked Caliph Umar what he should do with the books. "Throw the books in the water (meaning the Tigris River); if they contain any word of wisdom, they are worthless since we have the God's book (meaning Quran); and if they are profane, there is no place for them; the God's book is all we need," he replied (Reza et al., 1974, Iranian calendar, 1350). Thus, no Persian books of Achaemenids and Sasanids eras had survived.

Also, many of the later books were lost by Sultan Mahmud Ghaznavi (Mahmud Ghazni) of Ghazni (now in Afghanistan) who ruled from AD 998 to 1030, decades after Abbasid's Caliphs. It has been said that upon invading Neyshabur (or Neyshapur),[16] he took books from the library and burned them. Also, after conquering the City of Ray (or Rey), then a center of science and scholars in Iran, Mahmud removed the books from the library and threw them on fire (Abu Rayhan Biruni). Likewise, due to invasion and destruction by Turco-Mongols (AD 1336–1405) more books were lost in Iran. Thus, only a fraction of Persian writings from the tenth through thirteenth century have survived.

1.7 PERSIAN GOLDEN ERA DURING THE MIDDLE AGES

1.7.1 DARK AGES DURING CALIPHATE

It took Arabs nearly 20 years from 633 to 651 to conquer Iran. By AD 652, most of the urban centers in Persian lands, with the exception of the Caspian Provinces Tabaristan (now Gilan) and Transoxiana (old Turan),[17] had been dominated by Arab armies. They reigned Persia for nearly three centuries through early ninth century AD and forced people to speak and write in the Arabic language and to convert to Islam.

No one was allowed to speak in Persian in public or write in Persian. Indeed, Arabs killed many who spoke in his/her native language. So, Arabic became the official and national language for nearly three centuries from 652 to the death of the Abbasid Caliph al-Mugtadar in 932. During this Caliph, who ruled from AD 908 to 932 (295–320 AH), Abassid lost territory after territory; and at the end, had no dominance in Persia.

This, of course, was not incidental. The killing of Persian elites by Arabs resulted in the social, national and religious uprising of Persian people from the latter half of the seventh century. Thus, many Persian nationalists including Abu Moslem Khorasani, Hashem ibn Hakim, and in particular, Babak Khorramdin[18] and Yaghoub Lays Saffar fought Arabs hard and long to stop their domination. These battles weakened Abbasid influence. In fact, by AD 853 (249 AH) many parts of Persia were governed by local dynasties.

In the AD 930s (320s AH), the daylamites (Daylamian), who were descendants of Sasanian soldiers and long resisted Muslim conquest and subsequent Islamization of Persia, managed to gain control of much of the present-day Iran. Daylami Buyid dynasty ruled for over a century until the invasion by Seljuq Turks in AD 1069 (447 AH). Daylamian supported the scientific activity of Persian scholars including Abu Bakr Mohammad Karaji whose previously indicated treatise *Extraction of Hidden Waters* is the oldest book on hydrology. During the period of Arab domination, not only were Zoroastrian Persians forced to convert to Islam and Arabic was mandated to be the sole oral and literary language, but also Arabs attempted hard to change the Persian traditions and culture. While they were successful in changing the language for nearly three centuries, they failed to affect the culture and traditions. Arabs Islamized Persians, but they could not Arabize them; in fact, Persians created a new sect of Islam and branches thereof, different from that of Muhammad. Moreover, Persians gradually held secretarial (dabir) and administrator (tadbir) posts in Abbasid Caliphate, especially during the reign of Harun al Rashid from AD 786 to 809 (170–193 AH). These people, many of whom were imbued with the lure of Sasanian, taught Arabs certain Persian concepts in administration and polite society (Adab). Regardless, Arabs did not encourage any scholarly work other than medicine which helped their health and astrology for their fortune telling, and to a lesser extent, mathematics.

The period from the initial attack of Arabs in AD 633 (11–12 AH) to nearly the termination of their domination in AD 873 (259 AH) is the darkest era in the Persian history. During this 240-year period, Persians made little contribution to science and technology. It is to be noted that the first book of Arabic grammar was written by a Persian named Sibwayh, who was born in Hamedan in AD 760 (143 AH). After unjustly humiliated at a grammatical debate in Baghdad, he returned home in Shiraz and died of anger at a young age in AD 796.

1.7.2 MOST NOTABLE PERSIAN SCHOLARS OF THE MIDDLE AGES

Once the Abbasid influence in Persia ended, the Persian scholarly works began. During the late ninth through mid-twelfth centuries, the greatest Persian mathematicians, scientists, engineers, physicians, astronomers, chemists, and philosophers of all time flourished. Among notable of them are the following:

- Abu Jafar Mohammad ibn Musa Khwarazmi (AD 780–850/163–235 AH). Khwarazmi was born in Khwarazm, northeast of Persia, which was not under the influence of Abbasid at the time. He was a great mathematician, astronomer and geographer. Khwarazmi's *The Compendious Book on Calculus and Balancing* introduced the first systematic solution to linear and quadratic equations. The words *algebra*, derived from *al-jabar* and *Algorithm*, which is the Latin form of his name, reflect the importance of Khwarazmi's contributions to mathematics. Khwarazmi was a Zoroastrian at youth and had to convert to Islam before he worked as a scholar at the House of Wisdom in Baghdad between AD 813 and 833 (197–218 AH). A statue of Khwarazmi was installed at Amir Kabir University in Tehran. during Pahlavi kings.
- Abu Bakr Mohammad ibn Zakariya Razi. Razi was born in Rey (near Tehran) in AD 865 (251 AH). In their youth, he moved to Baghdad where he studied at the local hospital near the end of Abbasids rule. Later he was invited back to Rey by Mansur ibn Ishaq, a Samanian King who then ruled Rey. Razi served as the Chief of the hospital (bimarestan) there until his death in AD 925. He was a physician, alchemist, philosopher, and made fundamental and significant contributions to various fields and especially medicine. Razi is known as the greatest physician of all time. He was a pioneer in ophthalmology and the author of the first book on pediatrics. He made leading contributions in inorganic and organic chemistry and was the first person to produce alcohol and acids, among other chemicals and compounds. Razi also served as a teacher of medicine and was devoted to the service of his patients whether rich or poor. Some volumes of his works "Al-Mansuri" namely *On Surgery*, which he devoted to Mansur ibn Ishaq, and *A General Book on Therapy* became part of the medical curriculum in European universities. He has been described as a doctor's doctor, the father of pediatrics and a pioneer of ophthalmology.
- Mohammad Karaji (AD 953–1029): Abu Bakr Mohammad ibn Hasan Haseb Karaji was born in AD 953 (342 AH) in Karaj, a city near Tehran, and

lived early parts of his life in Rey. Upon mastering in mathematics, he was invited to move to Baghdad, which was then ruled by Buyids (Al-e Buya in Persian) dynasty. Buids were a Persian dynasty who had gained control of much of modern Iran by the 930s and later removed Abbasid from power in 946. These Persian rulers, unlike Arab Caliphs, encouraged scholarly work and supported scientists. There, Karaji collaborated with Fakhr al-Mulk, who was Vazir of Baha al-Daula (misspelled Dawla in English writing) and dedicated his famous book in mathematics titled *Al-Fakhri* to this vazir and pursuant to request by the same vazir, Karaji wrote the book *Inbat-al Miya al Khafia* (*Extraction of Hidden Waters*), which is the oldest treatise on the subject of groundwater hydrology. In different passages of this book, he also describes elements of hydrologic cycle and mentions that the earth is spherical in shape, though considers it to be the center of the universe several centuries before Johannes Kepler or Isaac Newton. Excerpts from this book, together with more information on Karaji's biography and his contributions to mathematics, are presented in the next chapter.

- Avicenna (Ibn Sina). Abu Ali Husayn ibn Abd Allah ibn Hassam (known as Abu Ali Sina or Ibn Sina in Iran) was born in AD 980 (370 AH) near Bukhara. At a young age he moved with his father to Bukhara, which was then a center of scientists and was ruled by Samanids. Ibn Sina was a great philosopher and physician and is regarded as one of the most significant thinkers and writers of the Islamic (in fact Persian) Golden Age. Of the 450 works he is known to have written, only 240 have survived including 150 on philosophy and 40 on medicine. His most famous works are *The Book of Healing*, a philosophical and scientific encyclopedia and *The Canon of Medicine*, a medical encyclopedia. The latter book served as a standard medical text until the mid-seventeenth century in Europe. His wisdom influenced Omar Khayyam and Shahab al Din Sohrevardi from Persia and many Arabs and Europeans. He died in AD 1037 (428 AH) at the age of 57 in Hamedan, Iran.

- Abu Rayhan Biruni: Abu Rayhan Mohammad ibn Ahmad Biruni was born in AD 973 in Kath Khwarazm, which was ruled by Samanids. He lived early parts of his life in Khwarazm and in 998 went to the court of Ziyarid,[19] Amir of Tabarestan, and wrote his first important work "al-Athar al Baqia" which literally means "the remaining traces of the past" and is translated as "Vestiges of the Past" or "Chronology of Ancient Nations." He later visited the court of Bavand (Bavandian, Persian rulers who were Zoroastrian and opposed Islam) ruler Marzuban (the fifteenth ruler of the Bavand dynasty from 979 to 986). In 1017 Mahmud of Ghazni or Ghazna (known as Mahmud Ghaznavi in Iran) invaded Rey and most scholars were taken to Ghazni, the capital of Ghaznavid dynasty which is now in central-eastern Afghanistan. Biruni was made court astrologer and accompanied Mahmud during his invasion of India. After a conflict with Mahmud, Biruni lived in India for a few years. He explored Hindu faith, learned Sanskrit and authored the book *Kitab Tarikh al Hind* (*History of India Book*), which became an encyclopedia of Indian culture for centuries.

Biruni was conversant in Arabic, Persian, Sanskrit, and also Greek. He was a pioneer in the study of all religions, Islam, Zoroastrianism, Christianity, Judaism, Buddhism, and other religions. Biruni was a mathematician, geographer, astronomer and natural scientist. He is called the father of geology. His calculated 6,339.9 km for the earth's radius is only 16.8 km (0.26%) less than the current value of 6,356.7 km. Biruni had a heated debate with Ibn Sina over the movement of the earth in which he repeatedly attacked Aristotle's celestial physics. He argues by simple experiment that contrary to Aristotle, vacuum must exist. Biruni did not explicitly state whether the earth was moving around the sun, but commented favorably on the earth's motion.

Biruni also presented the correct explanation of rainbow and the mechanism of artesian wells seven centuries before Antonio Vallisneri's[20] explanation of same mechanism in AD 1715. It has been said that Biruni had also designed some of the earth qanats in Khorasan. Biruni is regarded as the greatest scholar of the medieval Islamic Era (in fact, Persian Golden Era of Middle Ages). He is one of the most, if not the most, distinguished figures in Persian history. George Sarton, a famous historian, calls the first half of the eleventh century as the Biruni Era. And Eduard Sachau, a famous German scholar, describes Biruni as "The most thinker scientist that the history knows." Biruni died in Ghazna in 1048.

1.7.3 PERSIAN SCHOLARS BACKGROUND

None of the above-indicated scientists were Arabs and none, other than Zakariya Razi, worked for or associated with the Caliphate. Even Razi, the greatest Persian physician, spent most of his life in Rey and served as the chief of the hospital there by invitation of a Samanian (Samanids) King, a Persian who ruled Rey (also spelled Ray). Most of these scholars' writings were in Arabic which was the language people were forced to speak generation after generation; and therefore, had become the sole oral and literary language at the time, the same way that Latin was then the scientific language for Greeks and Romans.

These scholars were Muslims because the Islam religion was forced upon them. However, their religion was different from Muhammad's Islam and was invented by Persians to differentiate themselves from Arabs. Also, some of the Persian scholars of that era were superficially Muslims and did not practice the religion. Avicenna (Ibn Sina), for example, was known to be a heavy wine drinker. And Razi, the famous physician who lived during Abbasids, composed some papers about prophecy and denied its necessity, which is a pillar of Islamic religion (Nasr, 1969, p. 99).

The fact remains that the scientific contributions of all these great Persian scholars were supported by Persian rulers in Khwarazm and Tabarestan rather than Abbasid Caliphs and later by Daylamian in Baghdad when they took over Abbasids. Once Arab Caliphs rule in Baghdad ended Persian scholars, such as Mohammad Karaji, Ibn Sina, and Biruni, began their scientific work under the support of Samanids, Daylamids, and other Persian rulers. Of course, Jews and Arabs also made contributions during the Middle Ages. Arabs' contributions, however, were mostly related to Islamic philosophy and theology rather than mathematics, science and engineering. Even centuries later Ibn Khaldun (a Tunisian who lived from AD 1332 to 1406), who is one of the most famous Arab scholars, wrote about Islam philosophy and historiography.

The contribution of these Persians, contrary to what is termed the Islamic Golden Age by Arabs and westerners, is totally unrelated to the Islamic religion. The same way that any scientific contribution by Christians, Jews, and Muslims in the West is not associated with their belief. From the domination of Arabs through early eleventh century, the Persians wrote in Arabic, which had been mandated as the sole common and literary language in Persia. As reported in Tarikh Tabari,[21] when Arabs invaded Khwarazm, they did such a massacre of people who did not speak Arabic that for a long time, no one could write or speak in Khwarazmi. The fact remains that the most significant scholars of the late ninth through eleventh century were mostly Persians and this period was a Persian Golden Era. Moreover, there were no scientific Arabic words when Arabs invaded Persia in the seventh century AD. The scientific terms in Arabic were introduced by Persian mathematicians, engineers, scientists, and physicians during the Arabs' ruling. And though many Arabic terms have been replaced by Persian words following the termination of Arab Caliphs in the mid-tenth century, some Arabic words and terms still intermingle in the Farsi language.

1.7.4 Persian Contributions to Literature

Unfortunately, invasion of Persia by Mahmud of Ghazni, who lived from AD 971 to 1030, Ghenghis Khan of Temujin in China, who lived from 1162 to 1227 and their descendants who ruled through mid-fourteenth century, and then Timur the Lame,[22] known as Amir Timur, a Turco-Mongol military leader from Uzbekistan who lived from 1336 to 1405, retarded the scientific work in Persia. However, during these periods (eleventh–fourteenth centuries) Persia created many world-known figures in literature and in epic, lyric, spiritual, and mystic poetry. These poets believed (as is proven now) that poem is far more effective than prose in expressing thoughts and ideas to be absorbed by people. The most notable of these poets and writers are the following:

- Ferdousi Tusi (AD 940–1020), misspelled as Ferdowsi in English literature, was born in Tous Khorasan and lived part of his life during Mahmud of Ghazna, who ruled until 1030. Ferdousi was mistreated by Mahmud, lived a poor life and spent 33 years (from 977 to 1010) to compose *Shahnameh* (*The Book of Kings*), in Persian language, the largest epic poetry ever written by a single person. Because of this book, he is called the savior of the Persian language and the reviver of Zoroastrian Heritage.[23] In the preface of this book, he writes: I endured much suffering during this 30 years, but I revived Ajam's[24] heritage with this Parsi (meaning Farsi language. Ferdousi's statue and his mausoleum both of which were built in Tusi, Khorasan during Reza Shah Pahlavi in 1930s.
- Omar Khayyam (AD 1048–1131) was born in Neyshabur (also spelt Nishabur and Nishapur), Khorasan, and lived during the Khwarazmian, a Persian Empire. He was not only a poet, but also a scholar, polymath, mathematician, philosopher, and astronomer. In AD 1072, Khayyam documented the most accurate year length ever calculated by any astronomer.

He is widely considered as one of the most influential thinkers of the Middle Ages. One of his poems reads:

If I had my hands on the universe like God
I would remove this universe from the base
I would make a universe from anew
So everyone can readily enjoy a desirable life.

Omar Khayyam's mausoleum in Neyshabur was completed in 1963 during Mohammad Reza Shah Pahlavi, who was the second Pahlavi King from 1941 to 1979.

- Jalal ad Din Rumi (1207–1273), more popularly known as Rumi, was born in Balkh, which is now in Afghanistan. He lived during the Persian Khwarazmian Empire. He composed a poetry book named *Divane Shams e Tabrizi* or *Kulliyat e Shams*, which is a masterpiece of wisdom and spirituality. In this book, Rumi expresses his devotion to Attar Neyshaburi, another Persian poet by whom he was inspired. He writes:

Attar has travelled the seven cities of love, we are still at the turn of one alley.

- Saadi Shirazi (1210–1291 or 1292) was born in Shiraz, lived during Chinese Mongol rulers, and spent a good part of his life working abroad. He was also known as Saadi of Shiraz; and under his pen name Saadi, wrote two books: *Golestan* (*Flower Garden*), which combines poetry and prose and is full of short stories, mostly in the form of admonition, and *Bustan* (*The Orchard*), which is a poetry book. The former book has been translated to many languages, including English and German. As indicated earlier, one of his poems is the motto of the United Nations gate. Saadi's mausoleum which in Shiraz was built between 1950 and 1951, during the second Pahlavi King.
- Khawja Khajeh Hafez-e Shirazi (1315–1390) was born in Shiraz and lived the early part of his life during Mongols and the rest during Timur Lame (Lang). He is known by his pen name, Hafez; and his poetry book, named *Divan Hafez*, can be found in most homes in Iran and Farsi speaking Afghanistan. *Divan Hafez* is full of life and joy and drinking wine, which was forbidden by the rulers at the time. So the book targets the religious hypocrisy. One of his lines reads:

They closed the doors of taverns, God don't desire that they open doors of dishonesty and hypocrisy

Hafez's mausoleum which is in Shiraz was built in 1935during Reza Shah Pahlavi.

The religious rulers at the time of his death did not allow Hafez to be buried in a public Muslim cemetery.

1.7.5 SUMMARY OF PERSIAN GOLDEN ERA OF MIDDLE AGES

In short, what is inappropriately referred to as the Islamic Golden Age is, in fact, the Persian Golden Era of Middle Ages. The scientific, scholarly, medical, philosophical, and poetical contributions by Persians during this era were totally unrelated with any religion. The majority of scholarly work occurred not during Abbasid but rather when the Arab Caliphs' domination of Persia was ended. In fact, during Samanian Empire (819–892) and Buyid dynasty (Al-e Buye, in Persia, 934–1062)[25] who encouraged scholarly work, the flourishment of Persian scholars reached its climax. It is time that the Persians be given proper recognition. The Golden Age is that of Persians rather than Islamic. The contributions are anything but religious.

Iran did not have a stable government during the fourteenth and fifteenth centuries. Also, the religious dynasties of Safavid (Safavian) from AD 1501 to 1736 and Qajar from AD 1791 to 1925 did not promote, rather discouraged any writing other than theology. Thus, until the King Reza Pahlavi, the founder of Pahlavi Kings in early twentieth century, Iran made far less scientific contributions than those in her Golden Eras. Pahlavi Kings made attempts to preserve and revive the Persian heritage including the calendar. However, after the Islamic Revolution in 1979, Iran has experienced the largest brain drain in her history since the Arabs' invasion during the mid-seventh century AD.

NOTES

1 Robert William Boyle (1627–1691) was an Anglo Irish natural philosopher, chemist and physicist.
2 During this period, the Greek cultural influence was at its peak in Europe.
3 AH (Anno Hijri), Lunar calendar of majority of Islamic nations.
4 SH is Solar Hijra calendar. This calendar was adopted in Iran in 1925 under Reza Shah Pahlavi who was the Prime Minister at the time, and like Gregorian Calendar, has 356.2424 days.
5 The water curvature is due to the force of gravity, as we now know.
6 Abu Rayhan Biruni was from Khwarazm (northeast of Persia, now in Turkmenistan, Uzbekistan and Kazakhstan). He is regarded as one of the greatest scholars of the medieval Persian (inappropriately referred to as Islamic) era. Biruni also distinguished himself as a historian, chronologist and linguist.
7 Golestan (Flower Garden or Rose Garden) has been translated by numerous people including Sir Richard Burton to English in 1928 and by Rudolf Gelpke to German in 1968.
8 Saadi is also known as Shaykh Muslih al Din Saadi Shirazi, better known as his pen name Saadi of Shiraz. He was one of the major Persian poets and literary men of all time and is recognized for the quality of his writing and the depth of his moral and social thinking. One particular poem of Saadi, written eight centuries ago, became a motto and decorates the main gate of the United Nations building in New York City. This poem reads:

> All human beings are members of one frame
> Since all, at first, came from the same essence
> When time afflicts a limb with pain the other limbs at rest can not remain
> If you feel sorry not for other's misery a human being is no name for thee.

9 The origin and construction of qanats are presented in Chapters 3 and 4.
10 One such code reads: "If a nun opens a tavern, or enters a tavern to drink, then shall this woman be burned to death."
11 Ctesiphon along the Tigris River was established as the capital city of Parthians a few centuries earlier.

12 Kurosh's "Charter of the Rights of Nations" was inscribed on a clay cylinder in Akkadian cuneiform. This oldest declaration of Human Rights was discovered in AD 1878 in a site in Babylon. It is now kept in the British Museum. The text on the cylinder was translated into all the United Nations official languages in 1971.

13 Pasargadae in Fars was founded by Kurosh as his new capital city.

14 The author's name should spell as Pazvash (pure like). However, his name was inadvertently written as Pazwash in his educational documents in the United States and has since been kept unchanged.

15 Transoxiana, also spelled Transoxania, called Fararud (beyond the Amur River), is the ancient name for central Asia in Persian land which is now Uzbekistan, "Tajikistan, southwest Kazakhstan and southern Kyrgyzstan. This land is a fraction of the Persian writings from the seventh through the mid-fourteenth centuries AD have survived.

16 This city, called Tisphon in Iran, was located on the eastern bank of the Tigris and about 35 km southeast of present day Baghdad.

17 Nishapur and Bishapur were named after Shapur I of Sasanian Empire.

18 Termed as "Turan' in Shahnameh, Persian national epic by Ferdoosi Tousi. Babak was a Zoroastrian Persian nationalist who killed tens of thousands of Arab invaders. His wife, Banoo, accompanied her husband in hard and long fights with Arabs. Babak was deceived by a Persian traitor, named Afshin, who handed him in chains and shackles to Caliph Mustasim to be slaughtered by a sword. Because of religious regime of Qajar, Babak was not well known in his homeland until the nationalist Reza Shah Pahlavi took power in the twentieth century.

19 Ziyarid (Ziyarian) was a Persian dynasty of Gilaki origin that ruled Tabarestan from AD 930 to 1090.

20 Antonio Vallisneri (AD 1661–1730) was an Italian medical scientist, physician and naturalist and served as the President of Padua University.

21 Jarir Tabari was born in Amol, Tabarestan in AD 839 and wrote a comprehensive history book titled *History of Prophets and Kings* and died in 923 in Leiden, Netherlands.

22 Timur was lame in one leg from the injuries during a war. He is named Timur Lang (meaning lame) in Iran.

23 Two centuries before Ferdoosi Abbas Marvazi, a Persian poet wrote a poem book in Farsi (AD 809/193 AH).

24 Ajam is the word Arabs used for Persians.

25 The Buyids were descendants of Panah-Khosrau, a Zoroastrian from Daylam. For this reason, most historians agree that the Buyids were Daylamids.

2 Excerpts of Karaji's Book *Extraction of Hidden Waters* in the Early Eleventh Century

2.1 INTRODUCTION

Karaji's millennium book, titled which was written in Arabic (the official language of Persia at the time). The title of the book in Arabic reads: *Inbat al-Miyāh al-Khafiya*, which translates to *Extraction of Hidden Waters*. Karaji's name in full is Abu Bakr Mohammed ibn Hasan Haseb al Karaji, often referred to, in short, as Mohammed Karaji and seldom inappropriately as Hasan al-Hasib, see e.g. Wulff (1966). A brief biography of Karaji is included at the end of this chapter. The original manuscript remained unnoticed and hidden in a library for centuries. Years after the discovery of an original copy of the manuscript in a library in Alexandria, Egypt, it was translated by Hossein Khadiv Jam into Persian (Farsi) in 1966. This chapter includes excerpts from the Karaji's on the following subjects:

- Earth Description
- Force of Gravity
- Hydrologic Cycle
- Origin of Springs and Rivers
- Movement of Water
- Occurrence of Groundwater
- Types of Groundwater
- Indicators of Groundwater Occurrence
- Water Rights and Buffer of Qanat

In the book, Karaji also describes the instruments, some of which he had invented personally, in carrying out the construction of qanat. A review of the book reveals the profound and exacting technical presentation of the hydrologic cycle, the earth's geometry, groundwater flow, the source of streams and springs, and as importantly, the force of gravity. It is regrettable that due to the inaccessibility of the book, the ethnographic description, historical treatises on the subject of hydrology in general and groundwater, in particular, had totally neglected the name of Karaji as the first scientific contributor to these subjects (Pazwash and Mavrigian 1981). The fact remains that this book, which was discovered in the second half of the twentieth century and was translated to Persian in 1966, is the oldest treatise on hydrology and

 DOI: 10.1201/9781032659930-2

groundwater flow, and hence a historical jewel piece indeed (Pazwash and Mavrigian 1980). Reviews of the book by Needham (1971), Nadji and Voight (1972), Davis (1973), Zaghi and Finnemore (1973), and Pazwash and Mavrigian (1981) have served as the earliest introduction to this precious work of Karaji to the western world. Since then there have been numerous reviews and discussions on this book.

As Boyle (1968, p. 676) states: "Most of the available Arabic and Persian scientific manuscripts have not been read in modern times, much less studied, and those texts that have been published are to a great extent the result of chance encounters. The current general pictures may be altered significantly with the study of any additional text."

Karaji's book has been translated into French by Aly Mazaheri (1973) and, at least in part, into Italian, Hungarian and Russian. An effort by N. Zaghi and J. Finnemore to translate Karaji's book *Extraction of Hidden Waters* to English was initiated but ended in vain because of a lack of communication between the two men upon the graduation of the former from Stanford University in the 1980s. The sample excerpts of this book which are included in this chapter and in the chapter on Qanat construction illustrate the extent of knowledge inherited in Karaji's book. Thus, as Professor Davis (1973) notes, Karaji requires recognition as a pioneer in the area of groundwater hydrology. And as Pazwash and Mavrigian (1980–1981) had indicated, Karaji deserved a millennial celebration.

2.2 EARTH DESCRIPTION (KARAJI, PP. 3–14)

After an introductory remark, Karaji states that "Earth with all its mountains, plains, hills, and valleys has a spherical geometry. God has placed it at the center of the universe so that it continually turns around this center; however, it is a very tiny element of the living universe." This statement demonstrates that Persians knew that the earth is spherical in shape centuries before Galileo Galilei,[1] Johannes Kepler and Isaac Newton.

He continues: "The great God has created the earth with a solid center and no voids therein. And has placed each star, celestial body and the four elements: fire, air, water and soil,[2] at a specific place, such that when it is separated from that place, it would return to the same location. For this reason, heavy objects such as soil and water have a tendency to reach this center; and the heavier the object, the larger the attraction (pull) of reaching the center." The last statement will be emphasized in the next section.

"If one questions the earth's sphericity on the ground that high mountains, canyons and hills and valleys on the earth change the earth's shape from a true spherical geometry, and that if every element on the earth is attracted to the center so that the earth becomes a perfect sphere; then the earth has to be continually in motion like a water which seeks its own spherical shape. The answer to these questions is that all hills, mountains, valleys, and rolling plains on the earth are insignificant compared to the size of the earth. Moreover, the heaviness is in equilibrium on opposite sites of the earth. Because of solid mountains on its surface, the earth cannot be a true sphere. If the earth attains a perfectly spherical shape, it will be wasted because, if the water exceeds the capacity of cavities inside the earth, it would cover the entire earth surface. And if the water was insufficient to fill the cracks, it might drop so deep that it would be impossible to extract."

"Also, if the earth was uniformly made of soft and discrete soil particles, it could attain a true spherical shape and would also be wasted. To avoid this condition, the great God created mountains, hills and valleys and plains on the earth, and placed high mountains far away from the earth center such that the opposite weights are in balance, and the earth remains at rest. However, it must be noted that there is constant motion inside the earth and some of these motions (earthquakes, author's note) result in the fall and collapse of structures and deviation of objects from the vertical alignment. Likewise, mountains and hills gradually fall due to attraction to the earth's center and collapse. Also, there is a constant motion in the loose soil in the ground so that the soil particles attach together and become compact and hard. The largest motion is the movement and displacement of vast bodies of water and large rivers from one place to another. This results in a non-equilibrium of weighs on the earth; and to reach an equilibrium, the earth shakes, resulting in displacement of seas and formation or desolation of springs."[3]

This statement of Karaji on displacement of weights implies that the accelerated melting of Arctic ice caps due to global warming, which changes the balance of weights on the earth surface, will result in earth shaking (earthquake). The increased frequency of earthquakes in recent years may indeed be due to the displacement of weights on the earth surface due to global warming (author's note).

As far as this book is concerned, there is no need to discuss any subject other than water. The above indicated attraction requires the soil to be at the center and surrounded by water. If the earth was a perfect sphere and was so solid that water could not penetrate it and the distance between its center to its surface were perfectly equal everywhere, then a sphere of water would cover the sphere of soil like the egg white covering the yoke within it. Whether the water depth was thin or thick, the sphere of water would be parallel to that of earth; and hence water could not flow. The water would cover the entire earth surface at a uniform depth; and other than the sea mammals, there would exist no life on earth.

If the earth was a true sphere as described above and the distances from every point on its surface to its center were equal but that there were cracks uniformly inside the earth, the water would be in no more than the following three states:

Water would cover the entire earth as a single ocean; or water would just reach the surface and the earth would be entirely dry; or the water would be inside the earth at certain depth and had no place to go, in which case the water surface would be parallel to that of the earth.

In all these cases, the water would be stagnant. In the case that the water was at a uniform depth below the surface, it would be very difficult to extract water unless the water was at a depth reachable for extraction by bucket and windlass. I presented this discussion to illustrate the nature of water and to prove that the water moves in order to attain a spherical shape. And when that condition is reached, undoubtedly the water would no longer flow. The same is the case for objects and structures above the earth surface in that their falling and destruction is due to the same attraction of reaching the earth center and the sphericity of the earth (Karaji, p. 4–5).

2.3 ON THE FORCE OF GRAVITY (PP. 3–15)

A matter of prime importance that repeatedly appears in Karaji's book is the attraction of all objects, and particularly water, to earth center. The same attraction, which is now known as earth's force of gravity and is attributed to Newton in 1727, was known to Persians over seven centuries earlier. It has been reported that Newton, sitting under an apple tree and in deep thought, saw an apple separated from the tree and fell down on the ground. Undoubtedly, neither Newton nor other western scholars before him were aware that a Persian scholar had explicitly stated "Every object is pulled towards the earth's center" (Karaji's book, *Extraction of Hidden Water*, p. 11).

First, it is worth iterating the statement: "Heavy objects, such as soil and water, have a tendency to reach this (earth) center; and the heavier the object, the greater the attraction of reaching the center" (Karaji, p. 3).

Excerpts of Karaji's statements on this matter are as follows:

There can be no running or spouting water on the earth or inside the earth unless its source lies at a distance farther away from the earth center (higher elevation, author) than the point where it discharges or spouts on the ground. The source of water in springs can be no exception to this case. *(Karaji, p. 7)*

Some waters discharge naturally on the earth without any excavation. The source of these waters, as was indicated, is the water inside the earth voids which when it reaches a location on the earth closer to the earth center than where they originated, they discharge on the ground. And, if the water in its route inside the earth encounters a solid soil which extends to a mountain top which is closer to the earth center than where the water was originated (meaning lower elevation, author), it discharges as a permanent spring at that location. *(Karaji, p. 9)*

There are continual motions inside the earth, some of which result in collapse and destruction of buildings and cause objects to deviate from a vertical alignment. Likewise, is the case for hills and mountains that gradually fall and collapse in order to get closer to the earth center. *(Karaji, pp. 12–13)*

If a wall is erected around a natural spring which spouts without any excavation, the water impounds and rises there and covers the land which was previously not getting any water; this is because the source of the spring is located at a place which is higher than the spring outlet. *(Karaji, p. 40)*

I have heard that there exist large springs of fresh water in some islands. Undoubtedly, the source of these springs is not the seas that surround them; the reason being that the sea water is located at a lower level than the island and that the sea water is saline but the spring water is fresh (literally translates as "sweet" versus "saline" in Persian language, author's note). Rather the source and origin of these springs are from far away places that are higher than the spring surface. *(Karaji, p. 7)*

Referring to movement of groundwater Karaji states: "This water moves through cracks and voids from points farther from the earth center to points closer to that center" (meaning from higher elevations to lower elevations, author's note) (Karaji, p. 15).

Such that the distance from the earth center to its surface be unequal and water flow from points farther from the earth center to points closer to it. *(Karaji, p. 5)*

These sample statements exemplify the Persians' awareness of the force of gravity, which Karaji refers to as the earth's attraction to pull objects, and "attraction to center." Though Karaji did not explicitly define the proportionality factor between the force of attraction and the mass of the body, he indicated that the attraction is larger for a heavier object and based many of his assertions and conclusions on this principle which is now known as Newton's Law of gravitational force.

2.4 HYDROLOGIC CYCLE

On the transformation of water to vapor and vice versa, which is one of the main elements of hydrologic cycle, Karaji notes (Karaji p. 5–7):

With transformation of air (vapor) to water during cold days at places with cool climate and turning water to vapor during hot days and warm climatic places, the movement of water on earth continues. This transformation is very important in the livelihood of earth. *(Karaji, p. 5)*

On infiltration (another important element of hydrologic cycle),[4] Karaji writes:

- "A sign of wisdom of the great God is the creation of mountains at many locations on earth that are connected to one another and extend several kilometers in length and width. In between these mountains, there exist valleys and irregular gullies. During winter, vapor solidifies at these places and the heavy snowfall which accumulates through winter, lasts until summer. With the sun shining on these places during summer, the snows melt and become the source of water for springs, streams, wells and qanats." Water from the snow melt enters the veins and cracks which exist inside the earth, flows downward and creates the source of water at faraway places. This is because the great God created small and large veins inside the earth for water to fill. Also in the earth are solid rocks, porous rocks, hard soils and barriers, some of which are vertical, some horizontal and some inclined. Their function on earth is analogous to the flow of blood inside an animal's body. For this reason, water is more plentiful at places and less at others. Likewise, the groundwater is at shallow depth at some locations, but well below ground at others. Also, in locations there exists groundwater which never depletes."

On evaporation, Karaji writes:

- "The sun which shines from dawn till dusk takes the freshest and purest part of water and transform it into air (vapor).[5] It is for this reason that sea water is heavy and stale as the sun has taken freshness and purity of water from its surface over a long period of time. The fact that sailors extract fresh water from the sea bottom to drink evidences this statement." (Karaji, p. 16)

- "Thunderstorms occur because of the increase of vapor which makes the air heavy and a large portion of the vapor transforms to water." (Karaji, p. 18)

2.5 ORIGIN OF RIVERS, SPRINGS, AND GROUNDWATER

Karaji at several places in his book makes statements about the origin of rivers, springs, and groundwater. The following are examples of these statements (p. 7–10):

- "The source of large springs is the accumulated snows on the above described mountains, i.e., those mountains where the snow never disappears. Most of such mountains are located at high geographic latitude where fewer animals exist."
- Referring to melting of snow on mountains due to sunshine, Karaji states: "The water from snow melt becomes the source of springs, streams, qanats and wells; and this water also moves downward in cracks and veins inside the earth and becomes the source of water at faraway places" (Karaji, p. 16). This statement also indicates the knowledge held about the force of gravity.
- "There can be no spurting or fountain of water (now called Artesian, author) on earth unless its source lies at a distance farther from the earth center (higher elevation) than the point where it discharges. The source of spring can be nothing but this. Likewise, the source of rivers is the accumulated snow on the previously described mountains, namely those where snow never goes away and are located at geographically high latitude.[6] And at such places, there are fewer animals."
- "Water that percolates under the packed snow piles on valleys between mountains and also in porous or sandy soil enters groundwater bodies that move through veins in the soil or joins stagnant water inside the earth and replenishes the springs."
- "If a part of water which infiltrates the earth reaches a solid soil, it becomes stagnant. And if a conduit is opened above this barrier, water rises into it to the level of its pressure. This type of water is called entrapped water (now called confined groundwater, author's note). The percolation of water in the ground creates natural springs and fills the voids in the veins underground."
- "In between mountains there exist valleys and winding gullies. The vapor solidifies at these places during winter and the heavy snowfalls that accumulate through the winter last until summer. With the sun getting closer to these places and warms the weather during summer, the snow melts and becomes the source of water for springs, streams, wells and qanats."
- "The origin of all water inside the earth and the source of springs, rivers, streams and all water bodies that exist on the earth is the rain and snow. If the rain and snow cease to fall, water depletes and the earth will be eventually wasted."

These statements, and especially the last one, clearly demonstrate that Persians knew that rain and snow were the source of rivers and springs at least seven centuries before it became known to European scientists.

2.6 MOVEMENT OF GROUNDWATER (KARAJI, PP. 3–10)

"Since the great God wished that water be in constant motion and flow from one place to another so that earth contains water and soil, that terrestrial animals live on earth and aquatic animals in water, that irrigation be possible and fruits and variety of vegetables be produced, that minerals be formed, that life be continued, that all living needs including foods, drinks, clothing, variety of jewels and drugs be met; he created mountains, valleys, hills, ditches, sinks, rolling plains and a variety of rocks and soils such that the distance from the earth center to its surface be unequal and water flows from points farther from the center to points closer to it (meaning from higher to lower elevations, author's note). The areas far from the earth center form plains and continents and contain earthly animals and water moves from these areas toward lower places; and due to the transformation of air (vapor) to water during cold days and climatically cool places and turning water to air (meaning vapor) during hot season and warm climatic areas, the movement of water continues.

This exchange of water to vapor is very important in the livelihood of earth."[7]

The following statements further prove that the Persians knew about the movement of groundwater and the source of springs. Karaji states:

> I have heard that there exists large fresh water springs in some islands. Undoubtedly, the source and origin of these springs is not the water of the sea which surrounds them. This is because the water level at the sea is lower than the ground surface of the island. Further, the sea water is saline, but the spring water is fresh (sweet in Persian, author's note). Rather, these springs originate from distant places where the ground level is higher than the springs and their source is the snows which melt under sunshine and move underground from summer to winter. Likewise, large rivers originate from the abovementioned locations. What supports this statement is the increased flow of the rivers during spring, when the sun is closer to the earth and melts the snow provided their source is located north of the equator. However, if their source is located south of the equator, the increase in their flow occurs when the sun is closer to that location. This is the case for the Nile River, which originates south of the equator, flows north to Egypt and its discharge increases in the fall. Most of the water sources in the fertile land on the earth are located north of the equator where the air contains moisture and is heavy. And that air in these areas continuously transforms to water and becomes the source of rivers, springs and water inside the earth voids. A scholar has said that cold air (meaning moisture) inside the earth turns to water and forms the permanent source of qanats and prevents water from being stagnant. *(Karaji, p. 9)*

Karaji iterates the ideas of his predecessors (mostly Greek, author's note) on the origin of springs. However, he presents his understanding about their source without refuting their ideas. The following excerpts illustrate this point.

"Some scholars have said that the source of springs on mountains may be a lot of vapor in a large cavity inside the earth from which vapor moves upward, turns to water and flows downward from the mountain."

However, he states: "Some of waters, which naturally discharge from the ground without a need for any excavation, originate from the water in veins underground which reach the ground at a location closer to the earth center than its source. And, if the veins inside the earth are within a solid soil which extends to the mountain top which is closer to the earth center than where the water was originated, the water discharges as a permanent spring there".

"In a village named Kondeh near Saveh (a city in west central Iran, author's note), I observed a river flowing through a valley and the river water was fresh. In the middle of the river there was a rock outcrop with three holes squirting bitter water, drinking of which would cause diarrhea. Undoubtedly, the source and origin of water ejecting from the rock was not the river, rather the water that had entered the ground at a location far away from the rock and had changed its taste moving through the soil enroute to the rock".

"In one of the years that the Diyala River had high flow, dikes were erected along the river to prevent flooding of adjacent lands. As a result, water level rose in the river; and subsequently water climbed in the wells, spilled over and flooded several homes. While the water in the Diyala River was fresh, though muddy, the water was clear but saline in the wells. This occurred because of the absence of any barrier to stop the flow inside the ground. Such is also the case in Isfahan plain which has some resemblance to that of the Euphrates" (Karaji, p. 18).

These excerpts indicate that Karaji (and in fact Persians) had a genuine knowledge of the types of groundwater and its movement. In addition, these statements provide not only a correct explanation about the origin of springs and rivers, but also indicate that contrary to Greek philosophers, Persians were aware that water cannot move upward against the earth's attraction to its center. As was indicated, this phenomenon, which is now known as the force of gravity and is attributed to Newton's finding in 1727, was indeed known to Persians at least seven centuries earlier.

2.7   OCCURRENCE OF GROUNDWATER (KARAJI, PP. 15–18)

After the foregoing discussions we now assert that: "God created a permanent groundwater which is in constant motion inside the earth, as is the flow of blood inside a living body. According to our predecessors, this water does not increase or decrease with an increase or a decrease in precipitation because it is derived from condensation of air to water inside the earth." Instead of disputing this idea, he explains his notion of groundwater source as follows: "Groundwater fills most of the voids underground and parts of it joins other parts provided that no solid barrier is encountered. This water flows through underground cracks and cavities from points farther from the earth's center to points closer to it. Thus, groundwater, analogous to surface waters which are partly stagnant and partly flowing, is stagnant as seas at places and flowing as streams at others. Stagnant waters are mostly located under vast deserts and lowlands and can be tapped at definite depths. Snows which cover slopes of mountains and fill valleys and floodways and remain there until rays of sunshine approach a vertical line (i.e. toward the end of spring, the author) mostly replenish groundwater. Groundwaters under such plains are more abundant than those in other places. The said mountains and all the lands surrounding them serve as sources of these groundwaters provided that no barrier is encountered. In these mountains, the slopes facing north are more moist and wet than slopes facing the east and west and the driest slopes are those which face the south.[8] The sun which shines on these slopes from dawn till dusk takes the freshest and purest part of water and transforms it into vapor. It is for this reason that sea water is heavy and stale as the sun has taken the freshness and purities of water from its surface over a long period of time. The fact that sailors extract fresh water from the sea bottom to drink is an evidence of this statement. Procedure for extraction of such water is as follows: They use a lead jar with tiny holes at its base and a slim neck connected to a tube made of waxed fine leather through which water cannot penetrate. They fit a globe snugly in the jar neck and lower the jar into the sea and allow it to touch the bottom. Then they exert a pull on a chord fastened to the globe to suck the air in the jar to enter into the tube. Water seeps through the holes and into the jar. They lift the jar to the surface by a second cord attached to the jar's handle to find it full of freshwater."

"An evidence of the fact that sun extracts freshness and purity of water is that any water left in ditches, sinks, swamps or at the bed of shallow springs is not fresh."

"However, it is to be noted that water cannot exist everywhere within the earth because there are many barriers (impermeable layers) inside the earth in horizontal, vertical or inclined directions. As a result, permanent large or small springs are formed on the earth or permanent swamps appear on the surface, or the land remains dry with no sign of water except at large depths or earth surface behaves as a sieve that never gets saturated with water except during heavy storms. The occurrence of a spring in a vast plain surrounded by dry lands, in which no groundwater is found except at large depths, evidences the foregoing statement."

"What evidences the differences in soils and proves the existence of abundant barriers and impermeable layers inside the earth is the occurrence of dry deserts, on the one hand, and productive fertile lands such as Euphrates Plain[9] on the other. Due to soil porosity, its uniformity and the absence of any impermeable layer inside the earth, water is found everywhere in the said plain. The conditions there are such that water level in the wells rise with increase in the river flow and drop with decrease in the flow to the extent that water level in the wells attains the same level as the water surface in the river."

"When water is in proper quantity which neither exceeds an appropriate amount nor is it deficient, land remains reclaimed permanently and secured from catastrophes and calamities. However, if water exceeds the appropriate amount, it creates floods. And if it is deficient in quantity, reclamation diminishes and famine results".

"Further, rainstorms result when water vapor builds up, air becomes heavy and a large portion of these vapors turn into water. Likewise, dry winds and more fire, result from decrease in humidity and water vapor and lightening of air."

"Among types of waters is "trapped groundwater"[10] and that is a type of groundwater which results from falling rain percolating through earth pores until it is intercepted by a flat barrier and becomes stagnant. If a qanat is constructed at the base of this water body, it will extract water as much as in storage and ceases to flow once water is depleted".

2.8 INDICATORS OF GROUNDWATER OCCURRENCE

Karaji describes the type of soils, ground cover, color of mountains and a certain vegetation which are indicators of the occurrence of groundwater. These indicators are briefly presented as follows.

2.8.1 COLOR OF MOUNTAINS (KARAJI, PP. 19–20)

"Blackish and moist mountains with rocks covered with mud evidence the occurrence of groundwater. And, according to our predecessors, the amount of groundwater decreases progressively in greenish, yellowish and reddish mountains. Among the blackish mountains, those with soft and flat and broad layered rocks have more water than others."

"The larger the rock to soil ratio in a mountain, the smaller the amount of water beneath it. There is no water in isolated and small mountains, especially if the rock is hard and abundant. This is because the snow does not remain on the top of these mountains for long. Most of the mountain ranges with valleys which cover a large area avoid the snow from melting until spring and summer and contain plenty of water regardless of their color. And the flatter and broader the mountain top, the larger the amount of water. Those mountains that are covered with thick and healthy vegetation and fully grown trees which hide snow from heat and direct sun rays, contain plenty of water, especially in the foothills facing north"[11].

2.8.2 TYPES OF GROUND (PP. 20–21)

"The lands which are connected to the above indicated mountains have groundwater. If a number of plains are connected to these mountain ranges, the plain closer to the earth center contains more water and the water is located at shallower depth than others.[12] This is especially the case if there is plenty of voids and pores in the soil in that plain".

"The presence of moisture or fog or dew on the ground during morning is an indicator of the occurrence of water in the ground. And if from a valley or bed of a stream in between two mountains a sound similar to the wind whistle is heard, there is water beneath that land provided that there is vegetation cover and dew on the ground. Otherwise, that sound is that of the wind which as it enters the ground, breaks the empty pores and voids in the soil and create that noise".

"Near Nahavand there is a gravelly area on the hillside from which a whistling noise is heard followed by spurting water[13] which flows downhill. This case occurs two or three times a day and sometimes even more during a single day".

2.8.3 VEGETATION (KARAJI, PP. 21–25)

Karaji lists a number of plants including those that he believed to be indicators of groundwater occurrence and those reported by his predecessors some of which he takes an exception with. Those plants which he believed to be indicators of groundwater and those he took exception with are: the following:

"Chase tree, Nightshade, Sorrel, Nettle, Borage, Prickly, Wild rue, Wild carrot (Daucus Carota), Adiantum (Par Siaveshan, in Farsi), and Khar Shotor (Hedysarum, a deep-rooted plant which is literally translated to Camel Thorn in English). The presence of all the aforementioned plants, if not cultivated, evidences the occurrence of groundwater. He adds: "However, I do not believe that the Papyrus, Purslane, Willow tree, Colocynth, hollow or solid stander Reed and Bulrush are the indicators of groundwater occurrence. These plants grow only along running streams or stagnant, shallow water bodies." "Prickly and Sorrel grow in water bearing lands where the groundwater is stagnant (now called perched groundwater). They say this because these two plants grow on the side or top of mountains".

"Also, freshness of vegetation on the land surface evidences the existence of groundwater provided that the vegetation is neither planted nor is it irrigated. However, Khar Shotor roots penetrate deep in the ground to reach groundwater. I saw a large Wild rue (Esfand)[14] plant in Baghdad along the Diyala River grown below a Khar Shotor plant. With the rise of water in the river, Wild rue roots got broken and a piece of it fell into the river. Observing carefully, I saw that attached to the remains of Wild rue were the roots of K Shotor extending from there to the water surface in the river. A piece of these roots was more than fifteen zara[15] long".

"A truthful man told me that: 'I was digging a well in a land where Khar Shotors were grown. In a well shaft I saw a root of these plants which had penetrated fifty zara to reach water.' Karaji continues: If watermelons are planted in lands with wild vegetation, the best are those planted in the middle of Khar Shotor roots. To do this, they cut the root open in the ground and place a few watermelon seeds in there and cover them with soil. These watermelon plants grow better there than other places and many other plants may be grown in the same manner"[16].

2.8.4   Dry Mountains and Barren Lands (Karaji, pp. 23–24)

"White mountains are waterless. Likewise, isolated mountains are dry, especially if they are mostly rocky. Also, a land far away from moist mountains has little water. In such lands no water can be found except at large depths. The lands lacking any vegetation is also void of water."

"A land with chunky soil resembling clay clod has no water. Likewise, a land covered with thin slate in lateral and longitudinal direction, as if it has been paved with stone, has little water. Lands covered with coarse sand and gravel also have little water. Likewise, low lands, which are under direct sunshine, are poor in water."

"Saying that a land is dry or has little water, it is meant that its water is located deep underground. Obviously, if the excavation of a well is continued in a land, eventually we reach water provided that no obstacles are encountered on the way. The water in lands with low water is undesirable. The same is the waters located at extensive depths".

2.9   TYPES OF GROUNDWATER AND DIFFERENCES IN THEIR TASTE (KARAJI, PP. 25–30)

"There are three types of groundwater. The first type is the main[17] water in the ground which does not increase or diminish with increase or decrease in rainfall, and its condition and quantity change insignificantly with time. The quantity of this water, which is mostly attached to soil mass, is proportional to voids and cracks in the soil and it is not affected by lapse of time and variation in temperature. If there are neither solid barriers nor large caverns which retain the water in the ground, the water flows in all types of soil due to the natural downward pull (the force of gravity, author's note). And, this water may lie at shallow depth at some locations or deep at others. The flow and movement of this water is minimal and resembles that of seas on the earth's surface. Any qanat dug in such soil has constant continual water."

"The second type is the water which originates from the continual transformation of vapor to water in the ground. This water continues to move as long as there is vapor available to transform to water."

"The third type is the water which originates from rain and snow. Most of the reclamation on earth depends on this water as this water is the source of large rivers, springs and qanats".

"Unlike the water in the seas and stagnant ponds and marches, the taste of water inside the earth is invariable. The reason being that the sun takes the freshness of water on the ground (due to evaporation, author's note); and as a result, those waters become heavy and change their taste. However, the water underground is free of such effects. The warmth (mild temperature, author's note) of water in qanats originates from heavy flows, unless it is caused by rotten soil."

"One of the scholars had said: Warmness is the sign of good water and coldness signifies its illness. However, I have heard of hot springs with the water as hot as a bath house. This type of warmness is due to rotten soil. The most durable and livable springs, whether originating from shallow or deep depth, are those with water which is fresh (called sweet in Iran) and mild. Cold waters flowing in a qanat are not long lasting because they are derived from melting snows; except if they originate from permanent snows on the ground."

"If the water in a qanat originates from rain alone, that qanat dries up after spring. The reason being that such water does not issue from both the sides and bed of the qanat, rather it seeps from either one side, both sides or the ceiling of qanat. The long lasting and main waters are those that come from bed of the qanat; and if the bed of qanat is excavated, the water seeps from all around and the seepage increases with digging. The water in every dug well stands at the elevation of its source. For this reason, galleries are dug at the bottom of the well parallel to the source in order to increase the well water.[18] Such water is not the main source of water in the well, rather it is secondary" (Karaji, pp. 26–27).

"The freshest water is that of rain and snow, then is the water that flows in good soil or passes over sand and stone chips and there are no water plants in their passage. The waters that do not possess these characteristics change their taste due to soil and vegetation along their route. Algae and seaweed also change the water taste. Therefore, fresh, saline, bitter, sulfuric and oily water or those containing mercury and sulfur and waters which taste like tar, alum and other material are created inside the earth."

"Drinking some waters causes diarrhea or sickness. The taste of these waters varies due to the type of soil they travel through and the worst of which are those with their source located in hard soil or soil with low water. The abundance of water in a qanat improves its freshness with time. This is because the constant flow of water in a qanat freshens water. All hard waters cause illness and are not good for health."

"Some waters may poison the person who drinks it unless the person has grown up with that water and has become immune to it. The best waters are those which are easy to digest, infiltrate rapidly and warm up and cool down quickly. Any water with qualities other than these is not good and causes sickness. Also, any water which a person needs to take more than or less than his diet is undesirable. And, any water that changes color in contact with earthy objects or air is also bad. The saline water which solidifies next to air is not suitable for potable use. The source of such waters does not occur anywhere other than fine particulate soils. There is no need to discuss other types of water beyond those described herein" (Karaji, pp. 27–28).

2.9.1 Means of Differentiating Light, Heavy, Fresh and Unsafe Waters (Karaji, pp. 28–29)

"Any water that changes its color is not safe; and if it smells bad, it is rotten. Also, if a water does not taste good, it is unsafe. If a water cannot be differentiated by color, taste or smell but it warms and cools rather quickly, that water is healthy; and any water that has been stagnant in ditches and swamps, that water is rotten.

If you have two different fresh waters and want to pick the safer one, weight two equal samples of these waters, the lighter water is safer. Alternatively, fill two new identical clay jars with water, place each on a tripod and leave a glass or laminated vase under each. After several hours measure the seeped water from each jar. The one with a larger volume of seeped water is fresher and superior. Alternatively, place equal volumes of those waters in two identical vases and add equal amount of fine sand to each vase and allow the sand to settle. The water in the vase which the sand settles sooner is fresher and lighter. Note that the soil absorbs the taste and hardness of water in the same manner that water takes it taste and hardness from soil."

2.9.2 Purification of Poor-Quality Water (Karaji, p. 30)

"In connection with the foregoing presentation, I add that: 'If you add fine clean sand or soil in a container of saline or heavy water and leave it alone until the water becomes clear, the water loses part of its salinity and hardness.[19] Repeating this process improves the water quality. And, if you pour this water in a new clay jar and allow the water to seep from its bottom, it loses its salinity and hardness further."

2.10 WATER RIGHTS OF GHANAT AND WELLS

2.10.1 Religious Doctrine (Karaji, pp. 42–46)

Karaji iterates statements by some religious hierarchies and indicates discrepancies among them. These instructions which have no scientific bases are omitted from this book.

2.10.2   ANOTHER VIEW OF GHANAT BUFFER (KARAJI, PP. 46–49)

"We now discuss the qanat buffer in relation to differences in soil and displacement of water from one qanat to another. The buffer varies considerably due to differences in soil[20] and I express my thoughts to the extent of my knowledge on this matter."

"If a qanat is constructed in a land where the soil has uniform porosity and the soil type does not change in any direction and qanat water originates from rain and the nearby rivers, that qanat has no buffer. Such is the case in a land around the Diyala River in Iraq where the water level in the wells rises and drops with the river flow. Therefore, any land which is similar to that land has no buffer. Because water enters any qanat constructed in such lands from far and near and all sides, especially if the qanat is deep and the water seeps in from the sides rather than the bottom.

"However, the water in qanats which are constructed in vast plains and which are surrounded by heavily snow covered mountains, originates from beneath these mountains. Such qanats do not have a main source; their soil is discrete, soft and full of pores. If a qanat is constructed in such land at right angle to the mountain and is extended to the mountain foot and there is no other source along the qanat, its buffer is approximately 500 zara on either side. And if another qanat is constructed in this land parallel or nearly parallel to the first qanat, the distance between the two qanats must be 1,000 zara so that each qanat has 500 zara buffer provided that the beds of the two qanats are at the same level."

"If a qanat is constructed at the foot of the above described mountain and is laid parallel to the mountain, the entire distance between this qanat and the mountain is considered to be its buffer. However, the buffer on the other side of the qanat is minimal provided that a second qanat is constructed parallel to the first one and it is not deeper in the direction of the plain. Since the source of water in the first qanat is from the mountainside rather than the opposite direction, its owner cannot interfere with the second qanat. However, if the bottom of the second qanat is lower than the first one, the owner of the first qanat can stop the work of the second one. This is because the water which moves through soil pores beneath the mountain finds a lower escape route, dips down and flows in the second qanat and thereby reduces the flow of the first one."

"The buffer of a qanat in a hard soil is less than that of a soft soil and the harder the soil, the shorter is its buffer[21] and the minimum buffer is 40 zaraa. Evidently, the decision on the extent of the buffer must be made by a professional who has knowledge about soils. All we said about the determination of buffer extent is a matter of guessing and estimation. Because of vast differences of soil inside the earth, it is impossible to make a definite determination."

The following observation evidences the above statements:

"In a vast desert where no water could be found except at great depths, I came across a high area which contained a few wells with water at shallow depths. A trustworthy man told me that when digging a qanat in that desert, he encountered a rich water bearing strata and consequently his qanat had a large flow. If a second qanat is dug in this land far away from the first qanat but located in the direction of flow to that qanat, it will intercept the water bearing strata and undoubtedly will reduce the flow to the first qanat. Therefore, because of so much variation in the soil underground, it is impossible to define exactly the buffer of a qanat. No one but God knows about the galleries and the water inside the ground."

2.10.3 More on Ghanat Buffer (Karaji, pp. 46–51)

"After presenting all the religious instructions about buffers of qanat which (at the time, author's note) takes precedence over non-religious directives, and describing other subjects which were neither adequate nor were they general; I now, to the best of my knowledge, discuss the buffer as it relates to the differences in soils, Karaii notes.

"We know that if a qanat is constructed in a barren land and someone else wishes to construct another qanat for himself, the owner of the first qanat, according to the religious laws, cannot stop him provided that the two qanats are 1,000 zara (zara was a unit of distance, equal to two-hand spans of a person) apart and their mother wells are at the same elevation. However, due to differences in soil type, a second person may construct a qanat closer than the said buffer to the first one. To determine whether or not he can do so, I opine the following: First a trial well must be dug at the location of the second qanat to tap the water. If the water level at the test well was higher or lower than the water level at the first qanat, his owner has no right to stop construction of the second one. Because, if the bottom of the second qanat is higher than that of the first one, there will be no adverse effect to the first qanat. And, if the water level is lower, it is evident that the water of the first qanat does not originate from the location of the test well. However, if the water level at the test well stands at the same level of that of the first qanat, a line must be drawn from the test well to the mother well of the first qanat and a second test well is dug midway along this line to reach the water. If the water level at this well is significantly higher or lower than that of the first qanat, the construction of the second qanat will not have an adverse impact on the first qanat. The reason being that the water level in the second test well indicates that the second qanat will not draw water from the first qanat. However, if the water at the second well was at the same level as that of the first qanat, the second qanat will draw part of water from the first one. Therefore the owner of the first qanat has the right to stop the construction of another qanat within 1000 zara so that each qanat has 500 zara buffer which is the religious buffer of each qanat,"

"If a new qanat is constructed in a desert where someone has already constructed a qanat and a dispute arises between the two owners, the above described procedure involving test wells must be followed in resolving the dispute. And the water surface in the above descriptions mean a surface parallel to the horizontal one.

"If the water levels at the two qanats differ slightly, there is if the levels are considerably different, then no one has the right to encroach onto the buffer of the qanat owned by another person, meaning to erect a building or cultivate or conduct other activities in the buffer. The buffer of a ditch or stream is limited to its dirt mound or each side and this buffer varies with the size of the ditch or stream and at a maximum is seven zara".

APPENDIX 2A

KARAJI'S BIOGRAPHY

Abu Bakr Mohammad Karaji was a Persian scholar and mathematician of the mid tenth century to early eleventh century AD. While Karaji is also known as Karadji, Karagi, al-Karaji, his full name was Abu Bakr Mohammad ibn Hasen Haseb Karaji (Nadji and Voigt 1972, Helweg 1973, Pazwash and Mavrigian, 1980, 1981).[22]

Karaji (AD 953–1029) was born in Karaj, near the ancient City of Ray (also spelled as Rey). Karaji lived about 20 years before Avicenna (Abu Ali Sina, a Persian physician and philosopher, AD 980–1037), and Abu Rayhan Biruni[23] (a mathematician, scientist astronomer and sociologist, 973–1048) who lived a century after Zakariya Razi (the greatest Persian physician of all time, AD 854–925). At youth Karaji learned mathematics in Ray which was an ancient city near present Tehran and along the Silk Road. Upon mastering on this subject he moved to Baghdad and became a science star in that city. There he related with Fakhr al Mulk Mohammad ibn Ali ibn Khalaf, the Vazir (minister) of Baha al Daulah (a head of Al-e Buya dynasty who ruled Iraq from AD 1011 to 1017/402 to 407 AH) and devoted his famous mathematic book *Al Fahkri* to this vazir. It is not known for how much longer Karaji lived in Baghdad after this vazir was killed in 407 AH (Anno Hijri). However, after leaving this city, he returned to his homeland, but en route he lost the books he had written and got very depressed. Karaji lost any desire to write more until he met another Vazir named Abi Ghanem, who requested him to write a book about water inside the ground (groundwater) and means of its extraction. To address this request, Karaji wrote his book *Inbat al Miyah al Khafiyah*, excerpts of which are presented in this chapter and some other chapters in this book. In the preface of his book, Karaji greatly praised this Vazir and wished him health and safety and long life. Karaji in the preface also states: "Thus, I do not know any profession more beneficial than extraction of groundwater. Because through this work, the land reclaims and man's life becomes organized and plenty of profit is gained."

Karaji significantly expanded the principles of Algebra and Algorithms written by Mohammad ibn Musa Khwarazmi,[24] a famous Persian mathematician who lived two centuries before him. Although some of Karaji's work has been lost, he is undoubtedly one of the greatest mathematicians of the golden era of Persian science. Also, his book *Inbal al Miyah al Khafia* is indeed a great treasure and the first treatise on the subject of hydrology and groundwater. As was indicated in the Preface this book was translated into Persian by Hossein Khadiv Jam in 1966 (Persian year 1345 SH[25]) during Mohammad Reza Shah Pahlavi. According to Adel Anbuba, a Lebanese scholar and mathematician, Karaji is the author of twelve more books in mathematics and engineering. All these books were written in Arabic, which was then the scientific language in Persia.

It is to be noted that all of Persian books whether in mathematics, medicine, philosophy, science and astronomy, were written in Arabic from the eighth through early eleventh centuries. Arabic was the official and literary language in Persia, the same as Latin that was the scientific language all over Europe during the middle ages and early years of the Renaissance. Therefore, the significant contributions of Persian scholars, which inappropriately referred to the golden age of Islamic science by westerners and Arabs, have no relation to religion. Also, some Arab writers, such as Rashed, incorrectly refer to some Persian scholars as Arabs. The period from ninth through eleventh century AD was indeed the Golden Era of Persia.

Karaji was one of the greatest mathematicians of Persia during the latter half of the tenth and early part of the eleventh century AD. Karaji's books in mathematics include his famous book *Kitab Al Fakhri*[26] *al Kafi fel Hisab* (*Book of Comprehensive Calculus*) and *Al Badie fel Hisab* (*New Calculus*) in which Karaji attempted to free algebra from geometry. His books also include *Ilal Hisab al Jabar wa-l Mughableh'* (*Reasons for Calculus, Algebra and Collation*), and *Mokhtasar fel Hisab wa-l'Massaheh* (*Compendium of Calculus and Geometry*). These are Karaji's books of which copy or copies exist. For example, the only copy of the 'Ilal Hisab al Jabar al Mughableh' book in twenty some pages exists in Oxford, England and a copy of Mokhtasar fel Hisab va-l Massaheh is kept at the Library of Alexandria in Egypt.

Karaji has also written other books in mathematics and geometry which have been lost. These books include *Kitab fel Hisab al-Hind* (*About Indian Calculus*) which Karaji references in his *Al Badie fel Hisab* book and *Kitab fel Istigra* (*Book of Induction Reasoning*) to which Karaji refers in his *Al Fakhri* book. Likewise, Karaji's books titled *Nawadir al Ashkal* (*Rare Geometries*) and *Kitab al Dawr wa-l'Wasaya*, which are referenced at the end of *Kitab Al Fakhri* of which copies exist in Paris and Cairo. Another lost book of Karaji is *Al Madkhal fel Elm al Noojum* (*Introduction to the Science of Astronomy*), which is referenced in a book by a later Persian scholar. Other than the above captioned books, Karaji has also written a book titled *Kitab al Uqud al Abniya* (*Book of Bridges and Structures*), which covered topics such as construction of buildings, bridges, castles, and qanats. This lost book is referenced in a book by Shams al din Bokhari, a Persian mathematician who according to Khadiv Jam (1966) lived three centuries after Karaji, and passed away in 749 AH.[27]

Franz Woepcke (1826–1864), a German historian and Orientalist, was the first person who paid attention to Karaji's work. In 1853, he prepared a brief translation of *Al Fakhri* and wrote an introduction to this book. His work opened the eyes of other orientatists to the importance of Karaji. The book he translated was an original hand written copy or Al Fakhri, which was kept in Paris Library and now recorded as number 2459. In that book the name Karaji was misspelled as Karachi (the Persian letter "ز" appeared as "چ"). Consequently, to the mathematical world, Karaji was often incorrectly referred to as Al-Karachi (Boyer 1986; Smith 1958, and Taton 1963). As Pazwash and Mavrigian (1981) indicated, this error was due to etymological descriptions (as in the stresses of inflection within Persian letters).

Adolf Hochheim (1840–1898), another German Orientalist, prepared a commentary on *Al Kafi fel Hisab*, another book of Karaji, and translated this book to German in three volumes between 1878 and 1880.

Until 1933, the knowledge about Karaji's books was limited to the above indicated two books that were translated by Woepcke and Hohkheim. In 1934, an Italian Professor named Girogio Levi Della Vida published a paper on importance of Karaji in Oriental Research Magazine and verified that the correct name of this scholar was

Karaji and that he was from Karaj near Ray, then a large city and home of scientists and scholars in Persia. There are also a number of thirteenth and fourteenth century books by Persians which reference Karaji's book in algebra and mathematics and correctly spell the name of Karaji.

It is worthy to note that all the above named great Persian men lived and produced their work after Abassid domination in Iran had ended. They wrote in Arabic because it was a mandated language in Iran for nearly three hundred years. Also, achievements of these men and other Persian scholars during the ninth, tenth, and early eleventh century had nothing to do with their religion, as Persians made little contribution to science when Arabs reigned in Persia from mid seventh to near the end of the ninth century. Therefore, the Persians significant contribution to science is indeed a golden era of Persian science rather than the so-called Islamic Golden Age.

There are many written biographies some of which, however aim at the contribution of Karaji to mathematics.

NOTES

1 Galilei (1564–1642) was the first western astronomer and scientist from Italy who said that the earth is round, that it is not stagnant, rather it moves around the sun. For this heliocentric idea, he was sentenced to house arrest by the church for the rest of his life.

2 At the time, it was generally believed that everything was made of one or a combination of these four elements.

3 This statement is a very delicate scientific subject that has been studied in recent years. Mr. Rothee, a French· Professor (1969) and then the Director of International Association of Seismology and Underground Physics, based on his various observations, has claimed that the construction of large dams, which puts an artificial load on the earth's thin crust at the location of their lakes, results in earthquakes. He exemplifies several cases in India, South Africa, France and the United States where the construction of large dams has caused earthquakes. The 1962 earthquake in Buin Zahra (near Ghazvin) which occurred after the construction of Sefid Rud Dam in Iran might have been related to this natural phenomenon. Adverse environmental effects of large dams are also discussed by Goldsmith and Hildyard (1984).

4 This element was unknown to western scholars until the seventeenth century. Even as late as 1921, Ototsky a Russian denied deep infiltration other than exceptional cases and limited quantities (Meinzer, 1934).

5 Then vapor was considered as 'air', one of the four main elements of matter.

6 Referring to places closer to the North Pole (author).

7 The exchange of vapor to water and vice versa is an integral element of hydrologic cycle, as known today.

8 This is definitely the case for the Alborz Mountains southerly of the Caspian Sea in Iran. Due to orographic precipitation, the area north of the mountain receives far more precipitation than the area to the south.

9 A part of present Iraq.

10 So called "perched groundwater."

11 This is indeed the case for the foothills of Alborz Mountains along the Caspian Sea which receive from 1,000 to 1,800 mm of precipitation annually, partly in the form of snow.

12 This statement also implies the knowledge of gravity pull.

13 Called geysers.

14 Wildrue (named Esfand or Espand in Iran) is a plant that its seeds are burned to banish the Evil Eye. Its Latin name is Peganum Harmala. It has been found to have antibacterial and anti-protozoa activity.

15 Zera, an ancient unit of length measurement (according to Reza et al., 1974. this length was approximately 50 cm). However, zara was equal to 36 fingers width which is about 65–70 cm.

16 I heard from a friend of my father that he had seen watermelon plants grown by Khar Shotor in Baluchestan, a southwest Province in Iran, in the 1950s.

17 This water is absorbed to soil particles and is called soil moisture.

18 In the west, digging galleries to increase well oil was introduced by Leo Ranney, a Texan petroleum engineer, in 1920s (Linsley et al., 1972), ten centuries later than Karaji's time.

19 The sand bed has been used in water treatment plants for the past 100 years and the sand filters which have been used for removing pollutants from stormwater since the 1990s work on the same principle presented in Karaji's book.

20 Unlike religious status, Karaji correctly states that the buffer depends on the soil type.

21 This statement is factual and agrees with today's hydrogeological principals. We know that for a given drawdown, the radius of influence is shorter in hard soils (low permeability) than soft soils (high permeability). Author's note; (Pazwash, 1975, p. 315).

22 The term ibn means son of. At that time there was no last name; men were identified by adding the names of fathers and grandfather to their first name.

23 Abu Rayhan Biruni is the first person who explained the physical reason for the formation of rainbow seven centuries before the President of Pavada University in Italy reached the same finding. He also lived in India for a few years and wrote a book titled *Tahqiq mal al Hind* (Research about India) which became an Encyclopedia of Indian culture. Eduard Sachau, a famous German scholar in describing Biruni writes: "The most thinker scientist that the history knows."

24 Khwarazmi (AD 780–850) was born and lived in Khwarazm, northeast of Persia, and the term "algorithm" is attributed to him.

25 SH = Solar Hijri; began from Muhammad's migration from Mecca to Medina. The SH calendar, like Gregorian, is solar based and has 365.2424 days.

26 Kitab means book. This Arabic word is still used in Iran.

27 AH = Anno Hijri (or Anno Hegirae in Latin); Lunar Calendar since Prophet Muhammed migration (Hegira) from Mecca to Medina.

3 Origin of Ghanat
Its Spread in the World

3.1 GHANAT DESCRIPTION

Ghanat pronunciation of qanat (qanat[1] in Arabic meaning channel), known as karez, kariz in Farsi, is an underground conduit that is dug into the alluvial fan and extends from a mountain foot to playa skirt.[2] Through ghanats groundwater has been conveyed from rich aquifers under foothills to barren plains and deserts and has turned them into fertile lands and urban centers. To provide ventilation and other considerations to be discussed in the next chapter, Ghanat conduit is tapped by a number of shafts (vertical wells). See Figure 4.1 in Chapter 4.

The depth of shafts of a ghanat generally increases toward the most upstream shaft at the mountain foot. The depth of the deepest shaft (madar-chah, meaning mother well) generally ranges from 15 to 100 m and the length of ghanat gallery may be as short as 1 km or more than 50 km. The distance between shafts is commonly 10–30 m. However, at a hill crossing, the distance between shafts can be as far as 200 m. The ghanat from Mahan to Kerman (two cities in southeast Iran), for example, is approximately 29.5 km long (Smith 1953). The deepest mother well in Iran is reported to be 360 m deep and the longest ghanat is over 70 km long (Beckett 1953).

3.2 ORIGIN OF GHANAT

While water wells, aqueducts, and weirs are related to many ancient civilizations such as Chinese and Romans, ghanat is special to Persia. Ghanats are spread throughout the Persian Empire which extended from Nile to Indus during the Achaemenid (Hakhamaneshian) Empire twenty-five centuries ago. From the first century BC to the Sasanian Empire, the border between Parthian and Roman (later Byzantine) had been the Euphrates River. Later, during the Sasanian Empire (AD 224–632) just before the Arabs' invasion, the border was nearly the same as that of the Achaemenid period.

Up to the mid-twentieth century, there existed nearly fifty thousand series of ghanats in Iran with a total length of approximately 300,000 km, which is nearly equal to eight tenths of the distance between the earth and the moon (Pazwash 1982). As will be noted later in this chapter, there are many ghanats in Afghanistan and Iraq, which were then a part of the Persian Empire. The Persians also built ghanats in Armenia and Egypt which were then in their territory. During Sasanian, the Persians brought the ghanat construction technology from the Near East to as far as Turkestan and China to the east and to North Africa on the west. Later, Persians introduced the technology to Arabia, Syria, Algeria, Tunisia during Arabs' reign and to Sicily and to Spain in Western Europe during Roman times. This technology was later moved to South America by the Spaniards.

DOI: 10.1201/9781032659930-3

Many of the terms that are used throughout the world for ghanat, such as kariz, kaahriz, and karez is rooted in the Persian language (Farsi). In Syria, ghanat is called kenayat which may have been derived from the Persian word, kandan (digging). Ghanat is termed as foggaras in Algeria and Tunisia, śariz in Saudi Arabia and Yemen, fugarā in Iraq and falaj in Oman and the United Arab Emirates (Cressey 1958; Wulff 1966; Agarwal 1980).

As was indicated in Chapter 1, the Arabic language became the literary language in Iran and all books written by Persian scholars prior to early eleventh century, whether in medicine, science, philosophy, astronomy or engineering, were composed in Arabic. As such, the word ghanat changed to gghanat, and the ghanat kan (ghanat diggers) to moghani, and these terms are still used in many parts of Iran. However, the Persian terms kariz or ghanat and ghanat kan are still common in the Sistan and Baluchestan Province, a province southeast of Iran and Afghanistan, which was then a part of the Persian Empire.   In Pakistan and Central Asia, which were in Persian land during Sasanian and remained free of Arab invasion, the Persian word ghanat and in Azerbaijan (then a part of Persia), the Persian word kahriz/kehriz is common. In Balochistan province in Pakistan which has similar climate to that of Baluchistan in Iran, ghanat is called kariz.

During the Arab's domination which lasted over 220 years, Arabic became a mandated language in all of Iran. So all works of Persians during the eighth through early eleventh centuries AD were written in Arabic (the mandated official language in Iran at the time). One such book was *Extraction of Hidden Waters* written during late tenth, early eleventh century AD by Abu Bakr Mohammad ibn Hasen Haseb Karaji (AD 953–1029). Excerpts of this book relating to the source of rivers and streams and occurrence and movement of groundwater were presented in the previous chapter. The book also discusses measures for maintenance of ghanat and its protection from erosion, excerpts of which will be presented in the next chapter.

The existing records reveal that the construction of ghanat was a common practice in Persia over three thousand years ago (Butler 1933; Pazwash 1982; Tolman 1937; Wulff 1966). Only a few writers of science history, including Biswas (1970) and Singer et al. (1954, 1956) were under the impression that ghanat was first constructed in Armenia. This is not tenable in lieu of so many written records now available. In fact, the historical records, as will be noted below, indicate that ghanat construction was common in other parts of Persia well before the technology was brought to Armenia which was then a part of Persian Empire, anyway. Some of these records are presented as follows:

- Cyaxares the Great, the king of Medes who ruled from 625 to 585 BC, was born in Ecbatana (today's Hamedan) and selected that town as his capital city. By uniting most tribes of ancient Persia and conquering the neighboring territories, Cyaxares made the Median Empire a regional power. After his father, Phraotes, was killed in a battle against the Assyrians led by Ashurbanipal, the Scythians (a group of Persian nomads) overran Median. Cyaxares, taking revenge, killed the Scythian leader and proclaimed himself as the King of Median. Then he prepared for war against Assyria. Cyaxares reorganized the Median army and allied himself with

King Nabopolassar of Babylonia, a mutual enemy of Assyria. To formalize this alliance, Cyaxares daughter, Amytis, was married to Nabopolassar's son, Nebuchadnezzar. These allies overthrew the Assyrian Empire and conquered Nineveh in 612 BC. This marriage, as a historian puts it, resulted in a new system of irrigation, namely ghanat, which surpassed other methods of irrigation that were common in Babylonia. Through this method, groundwater was collected and conveyed to irrigate the garden and farms of the palace. Apparently, Nebuchadnezzar learned from Persians to build a ghanat to dedicate it as a gift to his wife so that she views a garden similar to those in Ecbatana and feels at home.

- An inscription left by King Sargon II, the King of Assyria (722–705 BC), claims that he learned the secret of tapping groundwater upon invading the city of Uhlu in the old mining country of Urartu around Lake Urmia, now northwest of Persia. He noticed that the people enjoyed rich vegetation in the absence of any river and found the reason to be ghanat. His son, King Sennacherib (705–681 BC), carried out a great irrigation work around Nineveh (an ancient city in upper Mesopotamia) and built a ghanat to supply water to Erbil (also spelled as Arbil and Irbil). This ghanat was 20 km long with shafts 45 m apart and dates back nearly a century before the ghanat in Armenia was built. As of the 1940s, this ghanat was still operational (MacFarden 1942).

- Translation of Egyptian transcriptions reveals that following the defeat of the Egyptian army by King Dariush I (spelled Darius in English literature), of Achaemenid Empire, Khenombiz (the royal architect), and Silakes (the admiral) constructed ghanats similar to those in Persia to bring waters to the Oasis of Kharga (Butler 1933). In recognition of this deed, Egyptians built a temple of Amun and bestowed upon Dariush the title of Pharaoh (the title of ancient Egyptian dynasty). Remnants of these ghanats that reportedly were functional as of 1970s have been investigated. It appears that the ghanats tapped the underground water at the Nile River and brought it to a low land 160 km away (Reza et al., 1974).

- There are more ghanats in Iran than any other country. This is evidenced by Figure 3.1, which shows a map of ghanats in the world and indicates that in Iran there are more or nearly the same number of ghanats as those in the rest of the world combined. In fact, a large portion of other ghanats were built either during Achaemenids and Sasanids or because of the spread of ghanat construction technology by Persians.

- Henri Goblot (1963), who explored the genesis of ghanat for the first time in his book *Ghanats, A Technique for Obtaining Water*, correctly indicates that the innovation of ghanat construction took place in the northwest of present-day Iran near present Turkey and later was introduced to the neighboring Zagros Mountains.

- As indicated by Cressey (1958), Persians brought the technology of ghanat construction to India and Chinese Turkestan. East of the present-day Iran boundary, including Afghanistan, Pakistan, Central Asia, and Chinese Turkestan (Sinkiang), as well as parts of present Iran including the Sistan

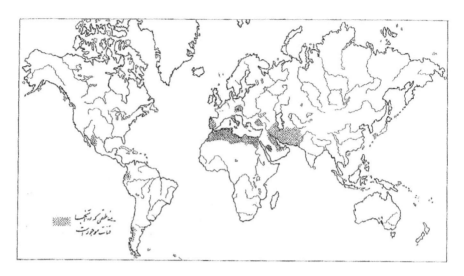

FIGURE 3.1   Distribution of ghanats in the world (Figure 90 in Reza et al.).

and Baluchestan Province, ghanats are known by their Persian term, kah-riz. In Sistan (now in Sistan and Baluchestan Province) where the Iranian Nationalist Yaghoub Lays Saffar, fought Arabs for many years, ruled Sistan, did not speak in Arabic and made Persian the official language there, the term ghanat, rather than gghanat is common. Also in Turkey, the gghanat is called ghanat. The use of Persian term ghanat, rather than the Arabic word *ghanat* or *gghanat* in these places which were not ruled by Arabs evidences that ghanat must have been invented in Persia. It appears that the work ghanat originates from *kuhriz* or *kuhreez* meaning water that seeps from mountain.

3.3   GHANATS IN IRAN

Ghanats are found in many parts of Iran, especially in the central plateau, the south-ern foothills of Alborz ranges and the arid northeast. In Varamin Plain (east of Tehran) alone, there were nearly two hundred series of ghanats, the largest of which used to deliver 500 L/s (Beaumont 1968). There were also four hundred ghanats in Yazd (a city in central Iran); and the city of Tehran alone had thirty-six ghanats, some of which were built over 230 years ago when Agha Mohammad Khan, the founder of Qajar dynasty, moved the capital city from Isfahan to Tehran near the end of the eighteenth century (Butler 1933; Clapp 1930).

These ghanats supplied the entire domestic needs of approximately 1.5 million inhabitants of the city until 1953 (Iranian calendar 1332)[3] when the water supply system of Tehran became operative. They also provided sufficient water to irrigate the gardens and farmland in the suburbs of the city.

Just before the completion of the water supply system the author remembers filling two clay jars from a ghanat, named Farmanfarma, two blocks away for potable use. A channel two doors away issuing from another ghanat provided water for watering vegetable gardens and flower beds and filling the pool at his family home.

Before the 1950s there were over forty thousand ghanats in Iran (Wulff 1966; Cressey 1958). With a total length of approximately 300,000 km, the ghanats were at least five times longer than all of the roads that existed in Iran in the 1970s. Reza et al. (1974) indicates that there were fifty thousand or so ghanats years before 1970. Also, Roohani, then the Minister of Water and Power, reported that there were fifty thousand karizes in 1967, of which thirty-five thousand were operative (Behnia 1988).

The majority of ghanats vary in length from 1 to 50 km. The depth of mother well of karizha is on the average 50–60 m. The shortest kariz is 200 m long in Zarand near Kerman. This ghanat has ten shafts with a mother well 15 m deep. The longest ghanat in Kerman is 29 km long with a mother well 96 m deep and has 966 shafts (Beaumont 1971). The oldest ghanat is in Zarach, a city in Yazd Province. This kariz is over three thousand years old, and 71 km long and is reportedly the longest ghanat in Iran and in the world (*Tehran Times Newspaper*, July 15 2016); though Butler (1933) refers to an 85 km long ghanat in Gharghiz Mountain in Isfahan Province as the longest one. The second oldest kariz in Iran, Ghasabeh ghanat, is in the Village of Gonabad, near Birjand in Khorasan Province. According to Noel (1944), this ghanat was still providing potable and agricultural water to nearly forty thousand people after 2,500 years. This ghanat is 33 km long; and with a mother well over 300 m deep, it is the deepest ghanat in the world. The Gonabad ghanat, which is also called Ghanat Kai Khosrow, was built during Cyrus the Great. This ghanat has 427 wells and is one of the eleven ghanats which are inscribed on the World Heritage Site by UNESCO in 2016 under the name of "The Persian Ghanat" (http://whc.unesco. org/ en/list/1506). Zarch Ghanat and Hasan Abad Moshir Ghanat (also in Yazd) and two other ghanats in Isfahan Province are also on the UNESCO list. The ghanats in Isfahan include Moun ghanat, being the only double decked ghanat in the world, and Vazan and Mozdabad ghanats each for using underground dams. Ebrahimabad ghanat in central Arak Province is also among the eleven UNESCO's registered ghanats for its conical shape of the conduit.

Discharge of ghanats varies from area to area and ranges from 10 to 50 L/s. The discharge also has a seasonal variation. In Mashhad area, a detailed survey by the Ministry of Water and Electricity included flow measurements of 249 ghanats for the two-year period 1343 and 1344 SH[4] (March 21, 1964 to March 20, 1966). The discharge of these ghanats showed cyclical fluctuations reaching maximum during spring season and minimum in autumn (Beaumont 1971).

Up to the middle of the twentieth century there were an estimated fifty thousand functional ghanats in Iran with a combined discharge of 750–1,000 m^3/s (Reza et el, 1974). This amount of water was sufficient to meet over 75% of domestic needs of then nearly 15 million inhabitants of Iran until 1950s (Noel 1944). It also provided water to irrigate over 2 million hectares of land. Since 1962, for the reasons to be discussed in a later chapter, many ghanats have been left in a state of disrepair and have dried up. Less than one-half of those remain functional as of 2017.

Yazd, Khorasan, and Kerman Provinces are known for their dependence on extensive system of ghanats. The city of Yazd in Yazd Province alone was watered by seventy ghanats, varying from 30 to 50 km in length with mother wells 50–125 m deep (British Admiralty 1945). The central and eastern regions of Iran contain the most ghanats due to low precipitation and lack of perennial streams. A smaller number of ghanats can be found in western Iran, which has permanent streams.

The annual amount of precipitation in the central plateau of Iran is generally less than 50 mm (Pazwash 1980). The scarcity of rainfall in this plateau makes traditional agriculture impractical. In fact, in other parts of the world, such as Central Australia with similar climatic conditions to that of central plateau of Iran, no agricultural activity takes place (Wulff 1968). Likewise, large portions of the States of Nevada and Arizona, due to insufficient precipitation, are left barren in spite of advanced irrigation technology in the United States. On this subject, William Marne (1960), then the director of the California Department of Agriculture, writes, "The Ghanat, ingenious horizontal well of ancient Persia which may hold promise for our water scarce southwest desert."

Through ghanat construction, which goes back to at least three thousand years ago, Persians could turn the deserts into fertile lands and agricultural plots. They built centers of civilization in desert like land which would otherwise be uninhabitable. A majority of the Persian ghanats bear characteristics to be called the feat of engineering, considering the intricate techniques used in their construction.

It was only due to ghanat construction that Sistan (a southeast province in Iran) with an annual precipitation of less than 50 mm (2 in.) could become one of the food producing centers of the world (Butler 1933). Or Yazd (a central city in Iran) could be built in a desert and become a center of civilization. Likewise, many large cities in Iran, such as Persepolis, Ray, Nahavand, Kashan, Kerman, Neyshabur, Mashhad, and Tehran, could be established only because of ghanats. Alas, some of these centers of cilivization were invaded and badly damaged due to invasions by Alexander and Mongols and more so by Arabs. Some of these cities, though were built again, did not achieve their past glory. As was indicted by Wulff (1966), "Persians have made a garden of what otherwise would have become an uninhabitable dessert." Figure 3.2 shows a Sabzevar ghanat which, as of the 1970s, was reportedly used as water supply system of this town in

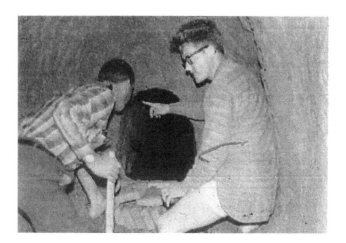

FIGURE 3.2   Sabzevar ghanat (Figure 69 in Reza et al.).

Khorasan Province. A description of ghanats in Sabzevar together with the role of ghanats as a sustainable irrigation system in central plateau of Iran are presented by Estaji and Raith (2016).

During the mid-1960s, based on Ghahraman (1958) estimates, nearly 15 million acres of cultivated land, one-third to one-half of the irrigated area of Iran, were watered by some thirty-seven thousand ghanats of which an estimated twenty-one thousand were fully operative and 16,500 though used were in need of repair. It appears that the 15 million acres estimate may be exaggerative. Based on an overall average annual irrigation demand of 650 mm/m² of land, the author estimates that with 750 m³/s supply of water, a net area of 2.6 million hectares (6.5 million acres) can be irrigated. This estimate is based on eight months of growing (cultivation) season, and 50 mm of precipitation during the same period.[5] Undoubtedly Ghahraman's estimate was not based on a 20,000 m³/s discharge originally suggested by Colonel Noel (1944). The 20,000 ft³/s (570 m³/s) would be a more realistic. Or perhaps his estimate relates to gross cultivation area, rather than the net irrigation area.

3.4   A HISTORICAL REVIEW OF GHANATS IN IRAN

The historical records indicate that the Achaemenid kings (550–330 BC) and Sasanid kings (AD 224–651) paid more attention to ghanat than Parthian kings who ruled from 238 BC to AD 224). For instance, Arsaces III (known as Arask or Arashk in Persian), one of the Parthian kings, destroyed some ghanats to halt the advancement of Seleucid Antiochus during a war. Also, during Achaemenid and Sasanid Empires, there existed a perfect regulation on both water distribution and farmlands. All water rights were recorded and the list of farmlands, whether private or governmental, was kept at the tax department. The government proceeded to repair or dredge the ghanats that had been abandoned or destroyed and even construct the new ghanats.

It was an Achaemenid ruling that anyone who succeeded in constructing a ghanat or in renovating an abandoned one and bringing water to irrigate land, he and his successors for five generations would be exempt from any tax to pay to the government. During Achaemenid Empire the technology of ghanat was in its heyday and it even spread to other countries including India, Turkestan and China (Cressey 1958) as well as Egypt.

During the Seleucid Era (311–238 BC), which began after the occupation of Iran by Skandar Maghdoni (known in the West as Alexander the Great), it seems that ghanat digging was abandoned and many ghanats were deserted and dried up.

During Arab Caliphs ruling in Iran, the ancient records of ghanat were ignored. According to the *Incidents of Abdollah ibn Tahir's Time*, written by Gardizi, a terrible earthquake struck the Town of Fergana in AD 830 and destroyed many homes. The inhabitants of Neyshabur kept coming to Abdollah ibn Tahir (the younger son of Tahir ibn Hussein) to intervene, for they fought over their ghanats and found no law on ghanats in the cleric's writings. So Abdollah ibn Tahir (who was the ruler of Khorasan under Abbasid Caliphs) gathered all the clergymen from throughout the Khorasan Province and Iraq to compile a book entitled *Al-Qanun* (*The Laws of Channel*). The book collected all of the clerical (Sharia) laws on ghanats which could be used to judge a dispute. Since there was no ghanats in deserts of Arabia, some of these laws were taken from Sasanian rules. A review of these rules is included in Chapter 2 on "Excerpts of Kariji's Book." Ms. Lambton quotes Moeen-al-din Esfarji's book *Rowzat al-Jannat* (*The Garden of Paradise*) that Abdollah ibn Tahir in the ninth century and later Ismael ibn Ahmad Samani (from the Samanid Empire) in early tenth century had several ghanats constructed in Neyshabur. Later in the eleventh century, Naser Khosrow (a writer) said that an Arab who was offended by the people of Neyshabur has complained that "what a beautiful city Neyshabur could have become if the ghanats flowed on the ground surface and instead its people would have been underground" (this evidences Persians hatred and rejection of Arab invaders and their continual fight to terminate Arabs ruling).

In the thirteenth century, the invasion of Iran by Genghis Khan of Mongolian tribes reduced many ghanats and irrigation channels to ruin. Later, during the Il-Khanid dynasty (AD 1231–1343) and especially at the time of Ghazan Khan and his Persian minister, Rashid al-Din Fazl-Allah, measures were taken to revive the ghanats and irrigation systems. During the Safavid Era (AD 1501–1736), the problem of water shortage intensified and led to the construction of many dams and ghanats. Jean Chardin, the French explorer who had two long journeys to Iran at the time, reports that "The Iranian rip the foothills in search of water, and when they find any by means of ghanats, they transfer this water to a distance of 50 or 60 km or sometimes further downstream. No nation in the world can compete with the Iranians in recovering and transferring groundwater. They make use of groundwater in irrigating their farmlands, and they construct ghanats almost everywhere and always succeed in extracting groundwater."

During the Qajar dynasty, who reigned Iran from AD 1796 to 1925, many ghanats were built in Iran, especially in Tehran. Thanks to Agha Mohammad Khan, the founder of Qajar dynasty who chose to move his capitol city from Sari to Tehran in 1776.[6] In move his capitol city from Sari to Tehran in 1776.[7] In the absence of any perennial stream or surface water, Tehran had to rely on ghanat. The rich supply of groundwater due to favorable geological and topographic conditions south of Alborz Mountains (Clapp 1930) allowed the construction of over thirty ghanats with a total discharge of approximately 2,000 L/s. Haj Mirza Aghasi, the minister of the Mohammad Shah, the third king of Qajar dynasty ruling between AD 1834 and 1848, encouraged and supported ghanat construction throughout the country. Jubert de Passa who surveyed the situation of irrigation in Iran, reported a population of fifty thousand in Hamedan, two hundred thousand in Isfahan, and 130,000 in Tehran in 1840; and claimed that the life in these cities were indebted to ghanat.

During the Pahlavi period (AD 1925–1979/1304–1358 SH), the process of ghanat construction and maintenance was continued until the Land Reform Act of 1962/1341 SH. Up to that time feudalism was the prevailing system in rural areas. Large landowners owned the wealthy lands and financed the maintenance of ghanat and irrigation systems. The peasants either worked for the landlord or paid rent for the land they used (Lambton 1953). The reform was in fact introduced by Hasan Arsanjani, a radical reformer who was then the Minister of Agriculture and who persuaded the Shah to put it to a referendum. There were over forty thousand ghanats with a total discharge of well over 600,000 L/s in 1942 and that the number of operational ghanat had dropped to thirty thousand with a total output of 560,000 L/s in 1961.

Dr. Mohammad Mosaddegh, who was the prime minster of Iran from 28 April 1951 to 16 July 1952 and again from 21 July 1952 to 19 August 1953 before he was imprisoned and put under house arrest after the 19 August 1953 (28 Mordad 1332) coup d'et at, had a practical plan to improve the life of peasants. According to his plan, the peasants would be freed from forced labor and 20% of the landlord rental income would be placed in a fund to pay for development projects such as rural housing, public schools, public baths and health clinics and pest control. However, this plan was not approved.

In 1962 a reform plan named the "White Revolution" was declared by Mohammad Reza Shah. The White Revolution was put to a national referendum and passed in January 1963. One of the articles of this plan included the Land Reform that let peasants take ownership of part of the landlord's land (Lambton 1969). Thus, agricultural land holdings were reduced from tens and hundreds of hectares (ha) to small plots, some as small as 0.5 ha. This reform was, apparently, intended to prevent an uprising and a revolt similar to those which resulted in takeover of communism in North Korea and Vietnam. The reform, however, ignored to consider the peasants' lack of financial ability to maintain ghanats and irrigation channels. In fact, this neglect was the basis of strong opposition to the land reform by the large land owners and ruling classes of Iranian society. Further, the reform neglected establishing an organization to substitute for the above indicated large land owners' function. As a result, ghanats were left in a state of disrepair and within just a few years many ghanats dried up and the peasants had to sell their land to agricultural corporations and individuals, and move to large cities, especially to Tehran in search of work. Lacking any experience in industrial work, many peasants were given low paying jobs and this gradually resulted in social unrest (Pazwash 1983).

In short, the land reform was a disaster to the country's agriculture. The country that had excess grains and produce just a decade earlier had to import grain to feed her own people. Those farmers who survived this upheaval and agricultural corporations drilled semi-deep or deep wells to irrigate their lands. Excessive draw-down of the water table accelerated drying up of ghanats, some of which vanished forever.

In March 1964, the Ministry of Water and Electricity was established in order to provide the rural and urban areas of the country with sufficient water and electricity. Three years later, the parliament passed a law protecting groundwater resources. This law allowed the Ministry to ban drilling any deep or semi-deep wells wherever the survey showed that they would result in depletion of groundwater.

This law and the law of "water naturalization" went into effect in 1968.

After the Islamic Revolution of 1979, attention was given to ghanats and the Law of Fair Distribution of Water was passed in 1981/1360 SH. Also, the first seminar on ghanats was held in Mashhad from June 27 to July 2, 1981 (6–11 Tirmah 1360 SH). During this seminar, representatives of various provinces reported the status of ghanats in their provinces. According to these reports ghanats supplied approximately 9 billion cubic meters (285 m^3/s) in 1980 which was three times less than the supply by deep wells. In many plateaus including Yazd, Kerman, and Mashhad, the water table had dropped due to boring of many deep wells and excessive withdrawal. In Yazd the drop during the 1955–1980 period was 20 m, which reflects an average drop of 0.8 m/year.

3.5 RECENT STATUS OF GHANATS IN IRAN

In 2000, a well attended International Conference on ghanats was held in Yaza. And in 2005 the Iranian government and UNESCO signed an agreement to set up the International Center on Ghanats and Historic Hydraulic Structures (ICQHS) under the auspices of UNESCO. The main mission of this center is the recognition, transfer of knowledge and experience, promotion of information and capacities relating to all aspects of ghanat technology and related historic hydraulic structures. The Gonabad ghanat was first added to UNESCO's list of Tentative World Heritage Sites in 2007 (1386 SH); and as was previously indicated, this ghanat with several other ghanats were officially inscribed in 2016 (1895 SH) under the World Heritage Site name of "The Persian Ghanat."

Since 1980 the population in Iran has climbed from nearly 38 to 88 million. Consequently, more and more deep wells had to be drilled to meet the water needs of the country. This has caused in continual drop in the water table and at places the wells have to be over 300 m deep to tap water. As a result, the ghanats have become a forgotten technology of suppling water.

As will be indicated in a later chapter, Iranian scientists performing rescue archeology at Farash historical site at the Seimareh Dam reservoir uncovered remnants of a water conduit in 2014. Iranian archeologists claimed it to be a water conduit from the late fourth millennium BC, and this turns the water conduit and irrigation technology some two thousand years back in time, over five thousand years ago.

3.6 OTHER USES OF GHANAT

In various parts of Iran, and in particular the eastern part of the country, there existed ghanats one behind another, each of which captured a portion of the water used for irrigation at the upper ghanat. Where the ground was steep, a 2± m water fall (similar to a drop manhole) was created at or near the ghanat outfall (mazhar) to run a mill. Such mills were used for grinding wheat or other grains. The water discharged from the mill would then be conveyed by a conduit and dropped at a second mill. In some places a number of mills were constructed along water falls of a single ghanat. The author had seen one such water mill when he was eleven years old. Ghanat outfalls are usually located at village squares and cultivated lands.

The outfall of a privately owned kariz was commonly at the garden or mansion of the owner. Some mansions were built alongside the kariz and a room (or rooms) was ventilated connecting the crown of the karriz to a hearth (like that of a fireplace) with a shaft. Through this practice the room was kept comfortably cool during summer and received some heat from the kariz water during winter. In a dry-hot summer climate, this could drop the room temperature by 16–20°C. A ghanat in a Manzarieh in a northern suburb of Terhran kept the rooms at a desirable temperature during summer and provided some heat during winter when the outdoor temperature dropped below 10°C (50°F).

During the summer of 1963, the author was on a practical training on Dez Dam (the tallest concrete arch dam in Iran) located near the City of Dezful. The room he had rented in a house was funneled to a ghanat at a hearth. This served as a natural air conditioner. To improve circulation of the cool air from the ghanat, a small electric fan was placed at the hearth. A wet towel placed on the fan helped improve the cooling effect even further. The room was thus kept comfortably cool while the outdoor temperature in midday rose to 54°C (130°F.). Besides cooling, ghanats have been used as wind towers in the deserts of Iran for over one thousand years (Bahadori 1978). Figure 3.3 depicts a schematic diagram of a typical wind tower.

3.7 GHANATS AROUND THE WORLD

A list of literature in ghanats around the world has been prepared by Wilke D. Schram. This list may be viewed at W.D.Shram@romanaqueducts.info.

A brief review of ghanats in various regions of the world is presented in the following subsections (see Reza et al,1974 among other references).

3.7.1 GHANATS IN THE MIDDLE EAST

3.7.1.1 Azerbaijan

Numerous ghanats (locally termed "kahriz") have been in existence in Azerbaijan (formerly part of Persia) for many centuries.

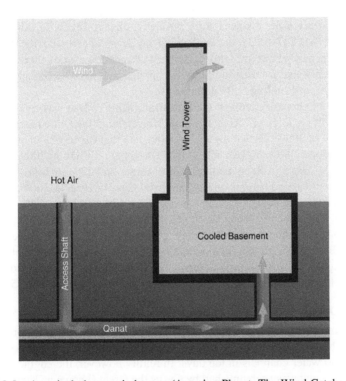

FIGURE 3.3 A typical ghanat wind tower (Amusing Planet, The Wind Catchers of Iran, February 2015, by Kaushik, www.amusing plant. com/2015/02/the-wind-catchers-of-iran. html) (Wikicommons).

Archaeological findings suggest that long before the ninth century AD, kahrizes brought potable and irrigation water to the settlements. Traditionally, kahrizes, meaning falling from Kah or Koh (Persian word for mountain) were built and maintained manually by a group of masons called "kankans." The profession, like in Iran, was handed down from father to son. It is estimated that until the twentieth century, of nearly 1,500 kahrizes, some four hundred were in use in Nakhchivan Autonomous Republic. However, following the introduction of electric and fuel pumped wells during the Soviet times, kahrizes were neglected. Today, some kahrizes are still functional in Azerbaijan and are key to the life of many communities.

3.7.1.2 Iraq

In Iraq, then part of the Persian Empire, there were many ghanats in the cities of Kirkuk, Arbil (also spelled as Erbil and Irbil), Sulaymaniyah (a city in Iraqi Kurdistan) and others. The city of Arbil, which used to be one of the largest and most populous centers of ancient civilization, could only be established because of ghanat. There were approximately 360 ghanats in the plains surrounding this city. Nearly sixty series of these ghanats used to supply from 20 to 30 L/s in 1955. Likewise, the ghanats in the City of Sulaymaniyah were providing the water needs of the city during the same year.

When Arab rulers learned about ghanat, they built them to bring water to their palaces. For example, Caliph Hisham in AD 728 brought water from Baghdad to his garden palace some distance away through a ghanat. Likewise, Abbasid Caliph Mutawakkil (AD 847–861) relied on Persian engineers to construct the water supply system for his newly built residence at Samarra. Recent excavation there has revealed that water was drawn from an aquifer of the upper Tigris and was brought to Samarra by ghanat systems totaling 500 km in length.

A survey of ghanat systems in the Kurdistan region of Iraq was conducted by the Department of Geology of Oklahoma State University in the United States on behalf of UNESCO in 2009. The survey findings indicate that out of 683 ghanat system in this region, some 380 were still active until 2004; but only 116 in 2009. Nearly 84% of the ghanats were in Sulaymaniyah Governorate and 13% in Erbil Governorate, especially on the broad plain in and around Erbil City (UNESCO, 2009). The reasons for this decline were neglect in maintenance, over pumping from wells, and drought since 2005. Consequently, water shortages in this region forced over 100,000 people, who depended on ghanat system for their livelihood, to leave their home. The study indicated that a single ghanat can potentially satisfy domestic demands of 9,000 people and irrigate over 200 hectares of farmland. This estimate, according to the author, appears to be exaggerated, unless the ghanat has a discharge of at least 25 L/s.

3.7.1.3 Oman

In Oman, once a part of the Persian Empire, there is a story that Solomon, the son of David, on his visit with a magic carpet ordered his djinns or community leaders to build 10,000 water channels in ten days. In reality however, the engineering feat of building a ghanat was accomplished by Cyrus the Great (known as Kourosh Bozorg in Iran), the Persian King of the Achaemenid Empire 559–530 BC (Trapasso 1996). Many ghanats, known as aflaj (the plural of falaj), were constructed; and the existing ones were maintained during prosperous times in Oman. During wars, alfaj were among the first targets of invading forces.

The gghanats created large scale agriculture in a dry environment. According to UNESCO, some 3,000 aflaj were still in use in the 2010s. Nizwa, the former capital city of Oman, was built around aflaj, some of which are still in use. In July 2006, five representative examples of this irrigation system were inscribed as a World Heritage Site (UNESCO 2009).

3.7.1.4   Jordan and Syria

Over 2,300 years ago, during Achaemenid Empire, water was brought to Petra (southern present-day Jordan) from Ain Musa, 8 km away, by an underground conduit. Through this conduit, which resembles the water conduit of Seimareh Reservoir, the people turned a dry desert into an oasis. The conduit consisted of vitrified clay pipes, each 1 m long and laid at approximately 4° slope. According to tests performed at the University of San Jose in California, if the pipes were laid at 6° slope, hydraulic jump would occur and the pipes would flow under pressure. However, at 4° grade, the flow in the pipe was fast but partly full. The conduit served Petra's population, estimated at 30,000 at the time. Nabataeans built five dams to control flash floods in Petra and 100 cisterns to collect the rain water that provided domestic demands of the city, estimated at 8 lpcd (liters per capita per day). Petra remained unknown to the western world until 1812 when it was explored by Johann Ludwig Burckhardt, a Swiss man. Petra was named the New Seventh Wonder of the World in 2007.

There exists a 94 km long Gadara Aqueduct, a Roman aqueduct, in northern Jordan; however, this aqueduct was never quite finished and was put in service only in sections. Cressey (1958) reports the existence of several ghanats in Jordan and Syria. In Syria, the widespread installation of deep wells has lowered the water tables. As a result, gghanats have gone dry and been abandoned across the country (http0://www.waterhistory.org/histories/ghanats/).

3.7.1.5   Arabian Peninsula

In the fifth century BC, the Persians introduced ghanat into Arabia (English 1968). According to George Cressey (1958), many of those ghanats were still operative then in Hejaz in the mountains of Yemen along the Hadhramaut, in Oman, and at Al Kharj Oasis southeast of Riyadh, and the Al Qatif Oasis north of Dhahran. Ghanats also exist in the Wadi Fatima west of the holy city of Mecca and Ain Zubaidah which bring water northwest to Mecca. According to the Persian historian Hamdollah Mostowfi, the gghanat in Mecca was constructed by Zubaidah Khatoon (wife of Harun al-Rashid, the fifth Abbasid's Caliph). And after the time of Harun al-Rashid (AD 786–809), this gghanat was decayed but was rehabilitated during the Caliph al-Muqtadir's reign (AD 929–932) and rehabilitated again after its collapse during the reign of two other caliphs, namely al-Qaim (AD 1031–1075) and al-Nasir li-Din Allah (AD 1180–1225). After the era of these caliphs, the gghanat completely fell into ruin due to filling up by desert sand, but later Amir Chupan (died AD 1327, a general of Mongol Empire) repaired the gghanat and made it functional again.

3.7.2 Ghanats in Africa

3.7.2.1 Egypt

As was indicated previously, during Darius the Great (Dariush Bozorg), the King of Achaemenid Empire in the fifth century BC, the Persians brought the ghanat technology to Egypt (Pazwash, 1982; Reza et al., 1974). The technology was also spread across the Fertile Crescent to the shore of the Mediterranean Sea,. In Egypt, ghanats that were built during the Persian occupation (525–332 BC) are found in Kharga Oasis and at Matruh. Beadnell (1909, 1933) measurement of one of the ghanats at Kharga, which were dug into soft sandstone, indicated that the ghanat was 3,200 m long and had 150 shafts. He also estimated that about 4,875 m³ of stone alone (7,300 tons) had been removed from the shaft and the conduit. Later, George Cressey (1958) found the remnants of the ghanat in Kharga Oasis. At the bottom of this ghanat, there existed an impermeable layer beneath the sandy soil which created natural springs.

3.7.2.2 Algeria

In Algeria, a number of oasis settlements had been irrigated until decades ago by ghanats (foggaras/fuggaras). In the Sahara region of Touat alone, there were several hundred km of foggaras which were reportedly functional as late as the 1930s (Butler 1933). The Tuareg who live on the southern fringe of the Sahara know the gghanats as "Persian work." The foggaras have been used since the eleventh century after Arabs took possession of the oasis and forced the residents to convert to Islam.

3.7.2.3 Libya

Ghanats, locally known as foggara, extend hundreds of kilometers in the Garamantes area near Germa, the ancient Capital of Libya. The conduits of foggara are generally very small, less than 60 cm wide and 150 m high, but some are several kilometers long and over 600 km in total length. The conduits were tapped by vertical shafts, one every 10 m or so apart, 100,000 m in total length and 10 m deep, but some reach 40 m depth.

3.7.2.4 Morocco

In parts of the Sahara belonging to Morocco, the isolated oases and Draa River[1] valley have relied on ghanat water since the late fourteenth century. There are a number of ghanats in the city of Rabat, which has geographic similarity to Tehran. In Rabat, ghanat is termed khettara. North of the city, the groundwater is not located deep underground, and as such, ghanats in this area are approximately 1 km long. The water temperature of these ghanats lies between 16°C and 18°C, and therefore, the water feels cool in the summer and warm in the winter. Until 1930, the old part of the city of Marrakesh and the European settlement part of the city, which is located at the foothill, were supplied only by ghanat water. In this city and Haouz Plains, the ghanats have been dried up and abandoned since the early 1970s. The impact of Hassan Addakhil Dam on local water tables has been one of the many reasons for the loss of half of 400 ghanats (khettara) in the Tafilalet Province.

3.7.3 Ghanats in Asia (Near and Far East)

3.7.3.1 Afghanistan and Pakistan

As was indicated, gghanat is known by its Persian term, ghanat or karez, east of present Iran and the Sistan and Baluchestan Province in Iran, as well. In Afghanistan, gghanats are called kariz and the gghanat conduit (Poshteh) is called Pashto. Thousands of ghanats exist in that country, which contains deserts and which used to be a part of Persia until 1857. Ghanats are the major source of irrigation water in the south and southeast Afghanistan, especially around the city of Kandahar (Humlum 1949). According to a report by the Ministry of Mines of that country, each ghanat had on the average 45–60 wells and irrigated 60 ha of land (Cressey 1958).

In Pakistan, the ghanat irrigation system exists only in Balochistan. The ghanats there were concentrated in the north and northwest along the Pakistan-Afghanistan border, which used to be in Persian land and the oasis of Makran Division. There, ghanats used to supply approximately two-thirds of the water in the city of Quetta and also irrigate some 35,000 ha (86,000 acres) of land in the vicinity.

3.7.3.2 India

There are ghanat systems in Gulbarga, Bidar and Burhanpur. The system in Bidar extends 2 km and has 21 shafts. This ghanat's vertical shafts are used by farmers and neighborhood settlements. The Indian Heritage Cities Network Foundation (IHCNF) managed by a board of trustees has been working toward conservation of the ghanat system. During a survey, IHCNF also discovered a royal bath ("Bagh-e hammam," a Persian word meaning "garden of bath") on the Bahmani period.

3.7.3.3 China

The Oasis of Turpan in the deserts of Xinjiang in northwest China had used water provided by ghanat, locally called karez. There are nearly 1,000 karez systems in Oasis of Turpan with a total length of about 5,000 km. Turpan has long been the center of a fertile oasis and important trade center along the Silk Road's northern route. The historical record of the karez go back to the Han dynasty. The Turpan Basin, with an average annual rainfall of 16 mm, had one of the most extensive karez systems in the world other than Iran until the eighteenth century. A karez in Turpan (or Turfan[1]) is considered one of the three great wonders of China, after the Han Dynasty Great Wall and Beijing-Hangzhou Grand Canal. As of the mid-twentieth century, approximately 40% of the people of this region depended on water of karez. Because of the importance of the Turpan karez water system to the history of the area, the Turpan Water Museum is a protected area of the People's Republic of China.

3.7.3.4 Japan

There are several dozen ghanat-like conduits in Japan known as "mambo," most notably in the Mic-and Gifu Prefectures in Chūbu region at the center of the country. Some link the origin of mambo to Chinese karez, and therefore, to the Iranian source.

3.7.4 GHANATS IN EUROPE

3.7.4.1 Spain

During the ruling of Umayyad Caliphate (AD 661–750), Persians spread the ghanat construction technology to Europe. Ghanats were constructed marginally in Catalonia Province in Spain and are still used in the Canary Islands. Sachau (1883) observed underground conduits with vertical shafts in the historic city of Palmyra and guessed these were ghanat. A few years later, Merchel (1889) verified this guess. In Spain, ghanat is termed as galleria. There are a number of ghanats in the city of Madrid. It appears that the term Madrid is the Latin translation of the Arabic world "Majra" meaning conduit. The Arabic term majra is still used to mean conduit in Iran. The City of Granada has an extensive ghanat system. Turrillas in Andalusia on the north facing slopes of Sierra Alhamilla also have evidence of a ghanat system.

3.7.4.2 Greece

There exists a number of ghanats in Greece and the island of Cyprus. The existence of ghanat in the Cyprus Island was first reported by an Englishman, Sir Baker (1879) and a century later by Cressey (1958). Cressy reported that the total flow of ghanat in Cyprus amounted to 9.25 billion gallons (35 million m^3) in 1950 and that more ghanats with an additional capacity of 1.85 billion gallons (7 × 10^6 m^3) were under construction. Goblot (1963), based on the reports he obtained from the officials in Cyprus, verified the flow of ghanats in that island.

3.7.4.3 Germany and Czechoslovakia

Ghanats were also discovered in Bohemia in the most western region of the Czech Republic and Bavaria in southeastern Germany. The first discovery of ghanat in Germany occurred in Selb, Bavaria in 1965. In a working day, the ground sank at Rosenthal china factory in Selb. When the sink was exposed, the workers found a water conduit underground. Following this incident, the Selb Research Institute was founded to investigate this underground conduit. It was found that local people, while aware of the existence of underground conduit, conceived it as an underground escape route. The Institute determined that this conduit was a ghanat, rather than a escapeway (Grewe 1998).

3.7.4.4 Italy

The Claudius Tunnel, intended to drain Fucine Lake (the largest Italian inland water) was constructed based on the ghanat technique. This 5,653 m long tunnel had shafts up to 122 m deep (Grewe 1998). Many of the ghanats are now mapped and some are open to the public for visitation. The famous Scirocco Rooms have an air conditioning system cooled by the flow of water in a ghanat and a "wind tower," a structure able to catch the wind and use it to draw the cooled air into the room.

3.7.5 SOUTH AMERICA

A number of ghanats exist in Mexico, Chile and Peru. There they are referred to as puquios or filtration galleries. The Spanish introduced ghanat into Mexico in

AD 1520. Ghanats also exist in the Nazca region at the southern coast of Peru and in northern Chile. In Chile, there are fifteen galleries near the city of Pica varying in length from 300 to 7,500 m and have a combined length of 30 km. As indicated previously, the Land Reform of 1962 in Iran disrupted the traditional feudalism agricultural structure in Iran. Traditionally, large landowners maintained the ghanats and irrigation systems and appointed bailiffs to supervise the allocation of water based on the size of the farm and the type of crop.

Through the Land Reform Act, the government bought the land from feudal landlords at a fair price and sold it to the peasants at 30% below the market value with a twenty-five-year loan payable at very low interest rate. The reform was meant to free 1.5 million families, some of whom were treated as servants little better than slaves, and to give them the land that they had been cultivating. Of course, there were some landlords who were fair to their peasants.

3.8 GHANAT, A LOST ART

As indicated previously, the Land Reform of 1962 in Iran disrupted the traditional feudalism agricultural structure in Iran. Traditionally, large landowners maintained the ghanats and irrigation systems and appointed bailiffs to supervise the allocation of water based on the size of the farm and the type of crop.

Through the Land Reform Act, the government bought the land from feudal landlords at a fair price and sold it to the peasants at 30% below the market value with a twenty-five-year loan payable at very low interest rate. The reform was meant to free 1.5 million families, some of whom were treated as servants little better than slaves, and to give them the land that they had been cultivating. Of course, there were some landlords who were fair to their peasants and helped them to improve their livelihood. Regardless, given the average peasant family size being five, the reform initially brought freedom to approximately 7.5 million people, one-third of the Iran population at the time.

The land reform on a short term achieved what it was intended to do. However, it neglected to account for the function of landlords as the provider of tools and goods to peasants; and more importantly, as the maintainers of ghanats and the irrigation channels. In short, the reform ignored that the ghanat ownership was directly linked to the land ownership. Before the reform, many of the ghanats were owned by a single family. However, after the reform, each ghanat was shared by a large number of landowners who neither had the financial ability to maintain ghanats and irrigation channels nor could they agree in paying a fair share of the maintenance cost.

Thus, within a few years, many of the ghanats were dried up due to the lack of maintenance, and consequently, more and more farmers steadily sold their lands to large private enterprises, abandoned the villages and moved to cities in search of factory jobs. This resulted in depopulation of villages and awkward growth of the cities (Pazwash 1983). Lacking any industrial education or experience, peasants were given low paying jobs, which resulted in a social unrest and vast depressed urban proletariat; a process that Fritz Schumacker referred to as "mutual poisoning." As was previously indicated, one of the intensions of the land reform was to avoid a Red Revolution (communism). However, it resulted in a general uprising and unrest;

and instead, it paved the way for the Islamic Revolution of 1979 (1357 SH). The new large landowners and agricultural organizations took advantage of the situation and exploited desperate peasants who did not want to leave. So, the land reform, which was initiated in the 1960s benefited only the new large landowners and large private business owners.

To supply irrigation and domestic water, these new owners had to turn to deep or semi-deep wells. As drilling by the new owners continued, the water table kept dropping which resulted in more and more ghanats to dry. At least fifty ghanats cited by name thus dried up in Kashan (a city in central Iran) alone and another thirty such incidents occurred in the city of Kerman southeast of Iran within a few years after the passing of the Land Reform Act. The impact was so severe that the ghanats discharge was reduced from 560 m^3/s in 1961–1962 to 380 m^3/s (1.8 billion m^3/year) in 1970 just a few years after the full implementation of the Land Reform Act and to 285 m^3/s in 1980. Thus, compared to a discharge of well over 560 m^3/s in 1961, the ghanats lost more than half of their productivity short of two decades.

A country whose surplus of grains and dry fruits were exported before the reform became so deficient in agriculture that it had to import a large portion of grain needs of her people. The fact that 458×10^3 metric tons of wheat were imported in 1970, and this figure the world has been prepared by Wilke D.

NOTES

1 In English literature, this word appears as "ghanat" or "kanat." However, in Iran, this word is pronounced "gghanat."
2 Until 1911, Iran followed the Anno Hijri (AH) calendar. On February 21, 1911, the Second Persian Parliament adopted Jalali (solar) calendar. During Reza Shah Pahlavi, the present Iranian calendar, which is Solar Hijri (Shamsi Hijri (SH)) and has 365.2422 days, was adopted on March 31, 1925 (the tenth day of 1304 Persian Year).
3 SH stands for Solar Hijri, which began from Muhammad's departure from Mecca to Medina in 622. Persia adopted the solar calendar in 1911, which was simplified during the late Reza Shah Pahlavi in 1925.
4 The author estimates that the cultivated land in Iran also consisted of 2.5 million hectares of rain fed irrigation, 0.5 million hectares of orchards and 1.5 million hectares of fallow irrigation at the time. In 2012, nearly 9.5 million hectares of land was irrigated in Iran; a large portion of which was irrigated by deep and semi-deep wells.
5 Tehran has since been the Capital of Iran.
6 Draa River is 1,100 km long; the longest river in Morocco.
7 In Persian language, the letter 'f ' and 'p' are, interchangeably, used in many words. It appears that the same was the case in the Turpan karez which was constructed by Persian moghanis or through Persian technology.

4 Construction of Ghanat

4.1 GHANAT STRUCTURE

Ghanat (qanat in Arabic[1], pronounced ghanta in Iran and Karaji's book kariz, or karez in Farsi) is a long, nearly horizontal underground conduit tapped with several vertical shafts. It extends a considerable distance from a mother well (mader chah) which dips down into the water bearing stratum (aquifer) to a playa. The mother well is dug at a foothill; where due to plenty of snowfall and heavy rainfall, the water table is steep and the groundwater is fresh. Through ghanat, which is laid at approximately 0.1% grade, fresh water is conveyed by gravity from mountain foot to arid and semi-arid plateaus to provide water for irrigation and to meet domestic demands of urban centers. For this reason, old civilizations that were established because of ghanats are referred to as ghanat civilization. Figure 4.1 depicts plan and profile of a ghanat system between the mountain foot and its mouth (outlet, called mazhar).

Ghanat may resemble a tunnel, however, its construction is far more involved than the tunnel technology. Construction of a ghanat requires a genuine knowledge of the occurrence of groundwater and its movement. For this reason, Tolman (1937), who is the founder of the groundwater hydrology in the United States, refers to ghanat as the most extraordinary work of ancient man. In fact, ghanat construction has been the most difficult, and laborious, yet very sophisticated and ingenious work of ancient civilizations. It is not only more wondrous than all other Wonders of the World, but it also has been more beneficial to man's wellbeing and

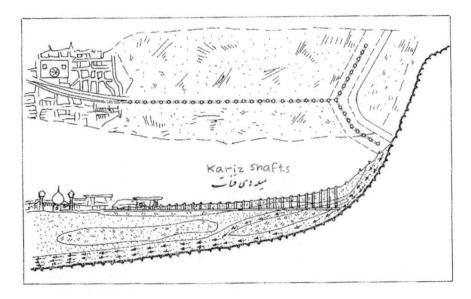

FIGURE 4.1 Plan and profile of a ghanat.

DOI: 10.1201/9781032659930-4

his prosperity than them all. The construction of ghanat, apart from its difficulties and hazards, is a very slow process and labor intensive. The author estimates that the construction of every 1 km of ghanat, which may include from 30 to 100 shafts, involves on the average 2,500 m³ of excavation and takes approximately four years for a crew of four. The excavation volume of a 3 km long ghanat with 90 m deep mother well in Khandab (in Zanjan province, author's note) which irrigated 8.7)ha and had included approximately 300 shafts, each 0.5 m² in cross sectional area was estimated at 85,000 m³ (Reza et al., 1974). However, this figure contains an extra zero; according to my estimation, the excavation volume would be approximately 8,500 m³. Working six days per week, con-structing a 10 km long ghanat with shafts 30 m deep on the average would take at least twenty -five years for a crew of six. Thus, building 300,000 km of ghanats in Iran must have consumed over 1.5 billion man-days. This amounts to engaging over 2,000 people during 3,000 years to build the ghanats in Iran. Such figure is thousands of time more than the labor to install 6,259 km of the China Wall above ground.

4.2   SEARCH FOR GHANAT LOCATION

Since construction of a ghanat requires a considerable capital cost and its return is unknown, the ghanat owner, whether a private land owner or a public entity, engages a professional to perform a preliminary study. The professional, who is usually an expert ghanat builder (moghani[2]), a class of professional well digger, carefully inspects the surface features of the ground where the ghanat is to be built looking for changes in soil cover, trace of seepage on the surface and hardly notice-able signs of vegetation; he searches where a test shaft (chah gamaneh, in Farsi) is to be dug. This location may be at the foot of a hillside or the bed of a dried stream. Moghani's experience and his knowledge of the water features of the area are his best guides for selecting a suitable ghanat location.

The following are excerpts of Karaji's text on suitable location for ghanat con-struction[3] (Karaji, pp. 106–108).

"In search for a suitable location to construct a ghanat, one should realize that the best place is the plateau of mountain foot which is humid and receives plenty of snow. A second best location is the land between the valleys of such mountains. A third good location is a plain that is connected to wet and broad mountains. When such a good place is found for ghanat construction, it should not be ignored. Also, the presence of plenty of healthy vegetation and greenery on a plateau that is far away from moun-tains is an indicator of rich groundwater and its suitability for ghanat construction; the reason being that the groundwater in such area is continual and independent of the quantity of the rain and snow on the mountain. Such knowledge will help selecting a suitable location for ghanat installation."

"When you find a suitable location, you start making measurements with a level from the mouth (point of discharge, called mazhar) upstream to the location of a test well. If the water level at the test well is higher than the ghanat mouth, you select that well as the starting point for the ghanat construction. The shafts dug downstream of the test well will be in dry soil and those dug upstream from it will be in water bearing strata which is the source of ghanat." "If it is possible to dig ghanat in hard soil, the digging should not be carried out in loose soil. Also, if in the course of digging you encounter odorous[4] soil in dry section, you should change the conduit alignment; past generations have advised us to avoid digging ghanat in bad odor soils."

"Where the soil at the ghanat conduit is not loose, the conduit can be made larger than normal. However, if the soil is loose and unstable, the conduit should be made smaller and its bottom should be curve shaped. A ghanat built in a soil where there exists longitudinal and lateral water bearing strata and seepage, that ghanat has high flow especially if the soil is dark."

"Ghanat construction may begin in the fall, namely late August through October (Iranian months of Mehr, Aban and Azara, seven through nine months after the first day of spring) when water is lower in the ground."

4.3   MEASURING DEVICES

4.3.1   LENGTH MEASUREMENT

For measuring horizontal distances along ghanat alignment on the ground, silk strings of 10, 30 and 100 zaraa (or zara)[5] were used. To avoid stretch, the string was carefully woven and waxed. Where more accuracy was needed, surveying chains, exactly 30 zara long were employed. Karaji in his book describes making the surveying chains as follows (Karaji, p. 73):

"The procedure for making a chain is to draw the copper in the form of a uniform wires and when the string is seen to be uniform, it is cut in pieces, each about hand span (22–25 cm) or somewhat longer. The end of each piece is bent to form a small ring and then these pieces are connected together. If the ends can be soldered, it should so be done; soldering is not difficult for a person who makes the string. This chain should be thirty zara long (Karaji, p. 73)."

4.3.2   VERTICAL RODS

In the old days, the Persians were using graduated wooden rods (Karaji refers to them as verticals). These were made 1.5, 2.0, and 2.25 m long. The 1.5 m wooden rods were used for leveling with water dripping levels or triangular plumb levels to be described later. The rods were legibly graduated so that they can be visible from a distance.

The 2.0 and 2.25 m rods were used for measuring elevation drop or rise in steep areas. To see the rod at a view point of the level, each rod was provided with a slider with a red dot at its center. This would indicate the difference in the ground elevation at the level and the rod location. A plumb bob was hung from the top of the rod to make sure that the rod was held in a vertical position. Figure 4.2 shows vertical rods, the longest of which is equipped with a slider.

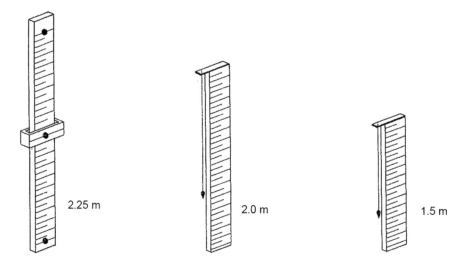

2.25 m

2.0 m

1.5 m

FIGURE 4.2   Vertical rods.

4.4   LEVELS

4.4.1   WATER TUBE LEVEL

This level is described by Karaji as follows (Karaji, pp. 64–65): "One of many devices used as a level is a tube made of perfectly straight glass or hard wood or bamboo and of uniform thickness."

The inner and outer surfaces must be smooth and parallel to each other. The tube inside opening must be the size of a small (pinky) finger and the tube's length should be about one and a half hand span. One hole is made at each end of the tube. A hole is also made at its middle for pouring water in the tube. These three holes must be along a straight line. You pick two strings, each approximately five zara long – zara equal to a forearm length. These strings should be made of silk or woven cotton and impregnated with wax so that it would not lengthen. Then connect each string to one of the holes in the tube."

Then two rodmen pull the strings from the opposite sides of the level. To keep the rod (vertical) in a vertical position as indicated, a plumb bob is suspended from the top of each rod. See Figure 4.2. Then the level holder pours a small amount of water from soaked cotton in the middle hole. When the water drips from the side holes at the same rate, the level is horizontal. A rodman holds the string at the top of his rod while the other man located at a higher elevation lowers the string slowly until the level lies horizontal, then he places a mark on the rod. This mark indicates the difference in the ground elevation at the locations of two rodmen." Figure 4.3 depicts the measuring process.

4.4.2   TRIANGULAR PLATE WITH PLUMB BOB

Another level is a board of isosceles triangle made of metal or hard wood that does not twist. A plumb bob is hung from the middle of the triangle base and when the

FIGURE 4.3 Water tube level measuring process.

base is held along a horizontal line, the plumb bob string will line up with the tip of the triangle. Figure 4.4 shows the triangular plumb bob level. It appears that because of ease of use and better accuracy, this level had faded out the water tube level.

After connecting the measuring strings to the holes at the base of the triangle, one of the rodmen pulls the string and keeps it at the top of his rod and the other rodman standing on the ground at higher elevation lowers the string until the plumb bob lines

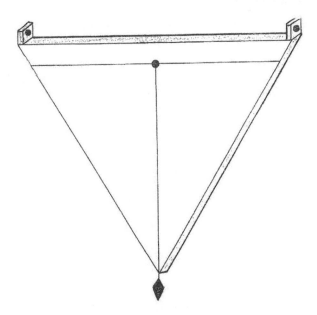

FIGURE 4.4 Triangular plumb bob level.

up with the vertex of the triangle. The readings on the graduated rod indicates the difference in the ground elevation at the locations of the two rodmen.

4.4.3 GRADUATED SCALE WITH PLUMB BOB

Karaji describes this scale as follows (Karaji, pp. 66–68): "Another level resembles a balance scale. This level consists of an arm and a graduated scale. The scale is made of thin steel sheet and to the extent possible light, but not so light that the scale is soft and can be easily bent. The length of the arm is approximately one and a half hand span and the scale is more or less the same length. From the arm of the scale a plumb bob weighing about 5 coins is hung from a thin string to keep the plumb bob in a vertical direction. Two hooks are fastened to the end of the arm and a string of certain length is connected to each hook. Alternatively, two holes are made at the ends of the arm for connecting the strings. Figure 4.5 shows this level."

"Then you pick identical wooden verticals having square cross section (square, author) each about 6 hand span long, with hanging plumb bobs and give them to two rodmen. One of the men holds the rod where the measurement begins and the other man holds the rod where the elevation is to be measured. Then the two rodmen hold the strings along a straight line at the top of the rod. The graduated scale on the base of the level directly indicates the difference in elevation at the location of the rods."

This device basically functions the same as the mechanical surveyor's (or miners' level) which were used up to decades ago before the digital and electronic levels were invented.

4.4.4 GRADUATED PLATE LEVEL

Figure 4.6 shows one of the levels invented by Karaji himself. Karaji describes the construction and graduation of this instrument as follows (Karaji, pp. 71–73):

> This level consists of a rectangular plate made of hard wood or brass with one hook attached at each side. You draw a straight line on the plate slightly below the hooks and make a hole exactly at the midpoint of this line. Then draw a line perpendicular to the first

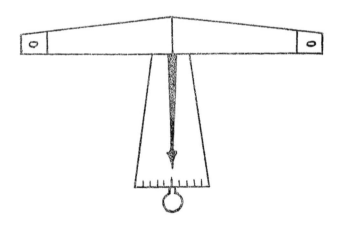

FIGURE 4.5   Graduated scale with plumb bob.

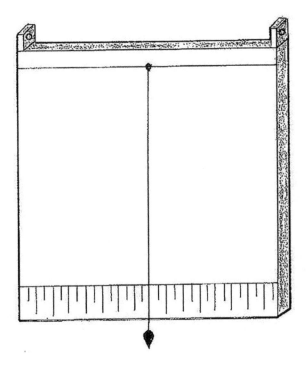

FIGURE 4.6   Rectangular level, invented by Karaji.

line and extend it to the end of the plate. And draw a line at the bottom of the plate exactly perpendicular to the line drawn from the hole. The last line will be parallel to the first line that you drew below the hooks. Then hang a plumb bob from the hole. When the last line is along a horizontal line, the plumb bob string forms two identical rectangles on the plate.

You pick two identical rods each graduated to sixty equal marks and two chains, each 30 zara long (or dhar was a measure of length and slightly longer than one meter) made of copper as previously described. Place the rods at two points at the same exact elevation opposite the rectangular level plate. Hanging the rectangular level from the two chains, the plumb bob string will coincide with the vertical line on the plate. Then you bring down the chain from one of the rods to the first mark on the rod while the chain on the other rod is kept at its top and place a mark at the alignment of the plumb bob. You repeat this process for each and every mark on the rod to prepare a graduated scale on the level. The finer the rod is graduated, the more accurate the level will be. After marking the level on one side of the plumb bob you repeat the process for the other side reversing the operation of the rodmen.

It must be noted that while the marks on the rod are equally spaced; the gradua-tion on the rectangular plate is non-uniform. However, each mark on the plate repre-sents a difference in elevation equal to the mark on the graduated rod. The reason for the marks on the level being unequally spaced is that the plumb bob string follows a curvilinear line. Karaji proves this matter brilliantly through a trigonometric method (Karaji, pp. 75–79, not included herein).

To use the instrument, two rodmen connect the 30 zaraa long chains to the hooks on the level, hold the chains at the top of the rod and the moghani reads the mark at the alignment of the plumb bob. The reading directly indicates the difference in the ground elevations at the locations of the rods which are 60 zaraa apart.

4.4.5 Rotating Tube Level

Karaji had also invented a tube level which would partly eliminate the use of measuring strings or chains or two rod holders. He describes this instrument as follows:

I have invented a level which is easier to use and is more accurate than other levels provided that its user is familiar with the device. This level consists of a round (or rectangular) plate made of brass or hard wood and a brass tube more or less one and one-half hand span long (35± cm, author's note). A small hole is made both at the center of the plate and at the middle of the tube and the tube is fastened to the plate by a hinge such that the tube can be rotated on the plate at the hole. The tube may be longer than the plate diameter. The plate is suspended from a string connected to a nail at the top of a hard wooden log approximately four hand spans long, so that when you sit on your feet, your eyes line up with the tube. If necessary, you can choose a longer or shorter log or lengthen or shorten the length of the hanging string to be comfortable in looking through the tube. Figure 4.7 shows this instrument.

Then you pick a square wooden rod of the same height as a man with raising hand, which is approximately nine hand spans. This rod must be hard, strong and perfectly straight. You mark sixty equally spaced lines on this rod and divide each spacing to the smallest practical segments. The upper and lower sections of this rod remain unmarked. At the upper most line a black, red or white dot, which would be visible from a distance, is marked on the rod. A similar dot is also placed at the lowest line on the rod. Then a rectangular bracket, which can tightly fit over the rod, is slid over the rod and a mark is placed at the middle of this bracket; see the left rod in Figure 4.2.

Level holder while looking through the hole in the level tube, has the rodman pull the string straight and slide the bracket on the rod to the point of his view and takes a reading. He then has the rodman to go behind him along the ghanat alignment until

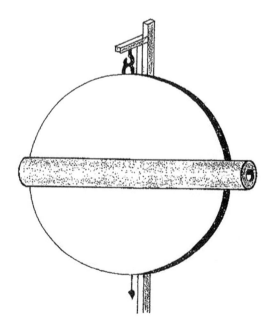

FIGURE 4.7 Rotating tube level.

the string is fully stretched, turns his level plate 180° towards the rodman, keeping the tube intact, and takes another reading on the rod. The difference in readings reflects the elevation drop (or rise) at these two locations. After this reading is done, he walks past the rodman along the ghanat alignment by a distance twice the length of the string, turns the level towards the rodman, takes a reading on the rod. Then he asks the rodman to go behind him again along the ghanat alignment and repeats the measurement. Thus, the ground elevations at points distanced twice as long as the string length are measured. Figure 4.8 depicts the measuring process using the tube level.

Karaji discusses two different procedures for measurement. One of these procedures involves using a string and the other a chain. The former procedure is described by Karaji as follows:

> Then he (moghani) picks a silk or tightly woven cotton string just thinner than a packing needle and connects its ends to two rings. The string should be about 100 zara long because eyes can easily see the dot on the rod from such distance. If the level holder can see farther, the string can be longer than 100 zara. One of the rings on the string is fitted over the graduated rod which is held by a rodman, while the other ring is placed on the rotating tube level. The level holder while looking through the hole in the level tube asks the rodman pull the string straight and slide the bracket on the rod to the point of his view and takes a reading. He then has the rodman to go behind him along the ghanat alignment until the string is fully stretched, turns his level plate 180° towards the rodman, keeping the tube intact, and takes another reading on the rod. The difference in readings reflects the elevation drop (or rise) at these two locations. After this reading is done, he walks past the rodman along the ghanat alignment by a distance twice the length of the string, turns the level towards the rodman, takes a reading on the rod. Then he asks the rodman to go behind him again along the ghanat alignment and repeats the measurement." Thus, the ground elevations at points distanced twice as long as the string length are measured. Figure 4.8 depicts the measuring process using the tube level.

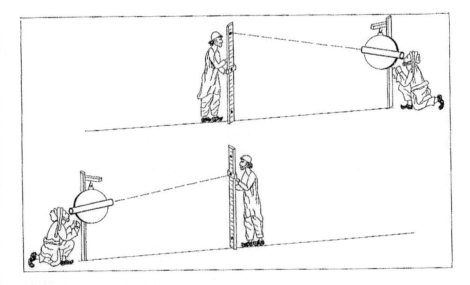

FIGURE 4.8   The measuring process using the tube level.

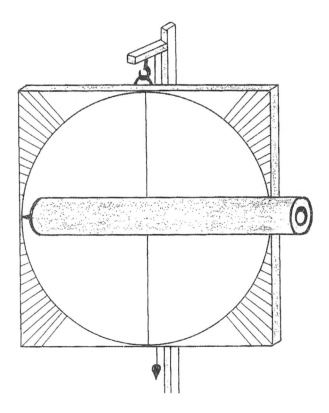

FIGURE 4.9   Rotating tube level with graduated plate for measuring high elevations.

Using the level for measuring the change in the ground elevation at two consecutive
shafts does not require using a measuring string. The level holder (commonly moghani)
places the level somewhere between the two shafts and takes a reading while the rodman is holding the graduated rod at the first shaft. Then he asks the rodman to move to
the second shaft, turns the level towards the rodman, looks through the same hole as
before and takes another reading. The difference in readings on the rod indicates the
change in the ground elevation at the two shafts.

Where the rise or drop in the ground elevation is more than the height of a measuring rod, a level consisting of a tube installed on a graduated plate, which shows the
angle of view, is used. Figure 4.9 shows this instrument. Karaji, in his book, describes
the procedure of using this type of instrument to measure the height of a mountain
without the need for measuring the distance from the instrument to the mountain foot.

4.5   GHANAT SHAFTS LOCATIONS

Using measurements with the level, the drop-in elevation between the trial shaft or
the head shaft (mother well) and the ghanat outlet (mazhar in Farsi), which may be
several kilometers away, is calculated. The survey measurements, which in old days
were performed by moghanis, also define the slope of ghanat conduit.

Depending on the depth of the trial shaft, the spacing between intermediate shafts is selected and their locations are marked on the ground. The depths of these shafts are calculated by a simple procedure as follows: The depth of the trial shaft where the ghanat digging begins is measured by a rope let down in the shaft. A mark is placed on the rope at the shaft level. Then, using a level, the difference in the ground elevation at the trial shaft and the next shaft to be dug is measured. Moghani then puts a second mark on the rope equal to the difference in the ground elevation at the two shafts. This mark will be below or above the first mark depending on whether the ground surface drops or rises between the two shafts. The length of the rope from its lower end to the second mark indicates the depth that the second shaft must be dug. This process is carried out for all shafts along the entire reach of the ghanat. The elevation drop between the bottom of two shafts is commonly neglected as the slope of the ghanat conduit is kept at 1 m/1,000–1,500 m (0.1%–0.07%). It is to be noted that the shafts may be as little as 10 m apart and within that distance the change in the elevations of the ghanat bottom is just about 1 cm; and therefore, would be insignificant. Where the shafts are far apart, the mark on the rope is adjusted to account for the slope of the ghanat conduit within that reach. However, this rarely happens.

The spacing between the shafts is selected based on the following parameters:

a. Aeration of ghanat conduit,
b. Overall excavation quantity,
c. Hauling of spoil in the conduit, and
d. Future maintenance.

It is evident that the further the distance from one shaft to the next, the smaller the excavation quantity, however, the longer the spoil has to be hauled in the conduit. Thus, near the ghanat outlet (mazhar) where the conduit is at shallow depth, the shafts may be 10 m apart, however, the distance between two consecutive shafts may be as far as 50 m or more approaching the mother shaft. In Sáadat Abad (a wealthy district in northerly suburb of Tehran) within the former property of the late Seyyed Ziaeddin Tabatabaee,[6] there is a ghanat with shafts about 200 m apart.

The chief moghani, based on his experience, selects the distance between two consecutive shafts so that the excavation is minimal while the ghanat aeration is not compromised. The objective of aeration is to allow the dampness to get out and fresh air to enter the ghanat conduit. Where the soil dampness is high, the distance between the shafts is reduced to improve aeration.

4.6 GHANAT CONSTRUCTION PROCEDURE

After selecting the trial shaft location, the ghanat builder, namely moghani and his crew, set up their windlass (charkh, in Farsi) on the upper slope of the ground. Then the moghani starts digging a well 80 cm (2.7 ft) in diameter using a broad edged pick (kolang, in Farsi) and a short-handled shovel. As the digging proceeds, the spoil is placed in a large bucket (delve) which is hooked to the windlass and is lifted up by two laborers who turn the windlass and empty the bucket around the mouth of the shaft. In old days, the buckets were made of untreated leather, but since early twentieth century, delves have been made of rubber. The mouth of the bucket which holds 25–30 kg of spoil is kept open by a strong circular iron loop.

As indicated, the trial shaft is sunk until it reaches a water bearing stratum (called ab-deh). The digging is then continued at a slower pace and a mixture of spoil and water (mud) is removed until the shaft reaches the bottom of the aquifer (zir su). By inspecting the mud lifted to the ground, moghani identifies the type of soil within the aquifer and defines the depth to the impermeable layer which is usually clay or sedimentary calciferous conglomerates. Upon completion of the test shaft, the quantity of inflowing water is noted for the next few days by emptying the well using buckets and leaving the water table to rise overnight. If the rise is measured to be 1 m or more from dusk until the next dawn, the aquifer discharge capacity is considered satisfactory. However, if the water entering the well is only from the soil moisture (earth sweat; aragh-e źamin), the rise is small, the groundwater is not genuine and the site is considered unsuitable for ghanat construction.

Where the test is satisfactory, another trial shaft is dug either upstream or downstream of the trial shaft; and by measuring the water level at the two shafts, the slope of the water table is determined. The deepest shaft with the highest yield and its bottom well above the ghanat mouth (mazhar) is chosen as head shaft or mother well (called "madar chah," in Iran). If the depth of the shaft exceeds 100 m or so, a second windlass is set up at a niche halfway down the shaft; and the dirt is transferred from one bucket to another at the niche.

After the test holes are dug, digging the conduit begins. The conduit must be dug along a straight line parallel to the alignment of the shafts on the ground. To dig the ghanat conduit (poshteh or koreh) extending straight along two consecutive shafts, moghani looks at a lamp placed at the center of the shaft where the poshteh digging had begun. The poshteh would be straight when he can see the light from both sides of poshteh. Seeing a dark shadow along the lamp indicates that the poshteh has not been dug along a straight line. Traditionally, the fat of pig or cow was used as the burning fluid for the lamp.

Karaji in his book (p. 108) states that

> I have personally observed that a number of moghanis who claim to be proficient had dug the ghanat conduit in dry section crooked and uneven. They had dug the bottom of the conduit either too high or too low and had created a dry bottom. And when they open the conduit to a shaft, they have to expand the sides to make the conduit straight. This process has significant adverse effect on the ghanat, especially if the soil is soft.

To dig a conduit straight in dry reach of ghanat, Karaji describes the following procedures (Karaji, pp. 109–115):

> The conduit in dry reaches of ghanat may be dug towards a given shaft or away from it. In the latter case, one can pick a wooden log about 3 zaraa long and four fingers by four fingers wide. Attach perpendicular to it a wooden board one and one-half zaraa long and hang a globe (plumb bob) from a fine string attached to the top of the vertical board. Then draw a straight line along the hanging plumb bob on the vertical board. Figure 4.10 depicts this instrument which resembles the inverted letter 'T'. Moghani can take this simple instrument with him and begin digging the first section of conduit from the bottom of the shaft. After digging one zeraa or so, he places the wooden 'T'

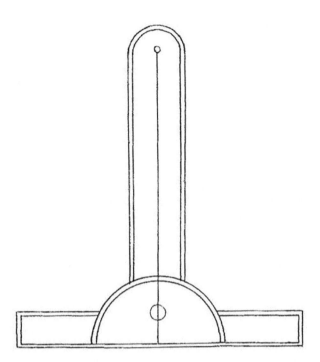

FIGURE 4.10   Wooden "T" and plumb bob instrument (not shown proportional).

upside down at the bottom of the shaft and the dug conduit section and looks at the plumb bob. If the string lines up with the marked line on the vertical board, the conduit has been dug flat (horizontal). However, where the string is inclined from the marked line, the bottom of the conduit is either too low if the inclination is to the right of the line and too high if the string is to the left of the line. The bottom of the conduit must be raised in the former case ad lowered in the latter case until the string coincides with the line. This process is repeated after the conduit is dug another zeraa or so until the conduit is dug straight and flat for 3 zara.

"To make sure that the conduit alignment is dug level beyond that point, a nail can be driven on the ceiling of the first section and a string is connected to it. And after digging a small reach of the conduit, the string is pulled along that reach. If the distance from the ceiling to the bottom of the last section is the same as that of the preceding section, the excavation has been carried out properly. However, if the distances differ, the excavation is improper. When a moghani finds that the distance is more than or less than that of previous section, he should raise or lower the bottom to correct the problem. Upon digging 3 zara or so, an experienced moghani can continue excavation correctly by viewing the alignment of the dug section."

Karaji also describes:

A more accurate instrument than the above indicated wooden T-log is as follows: Use a brass tube about one and a half hand span (35± cm, author's note) long and inside opening the size of small (pinky) finger. Attach a ring at each end of the tube and connect to the rings two fine identical chains, each approximately of equal length of the tube.

Attach the other end of the chains to a ring and hang the ring from a loop connected to a nail. Figure 4.11 depicts this instrument. The tube, when hung from the loop, must line up along a horizontal surface. To test this instrument, you can stand approximately 15 zara away from a wall, hang the instrument from a wooden stick, view the wall through the tube and have the point of your view marked on the wall. Then turn the tube 180 degrees and view the wall from the other end of the tube. If you see the marked point on the wall, the instrument is properly made; however, if this is not the case, the rings on the tube or the chains have to be adjusted to correct the problem.

"When you are done with the test and want to use the instrument to dig a straight ghanat in dry section, hang it from the center of the conduit ceiling after digging 1.5 zara or so. View the shaft where the conduit construction had begun and hang a small glove the size of chestnut from a string on the shaft wall. The globe should be exactly at the same height above the bottom of the shaft as that of the tube from the bottom of the conduit so that the globe can be viewed looking through the tube. As the digging is continued, hang the instrument from the ceiling near the end of the dug section and aim at the globe. If you see the globe through the tube, the conduit has been dug flat. Otherwise, you look through the tube from four corners of the conduit to determine in which direction the excavation has deviated. If the globe is above the sight, the conduit bottom is too low; and if the globe is below the sight, the bottom is too high. If the view of the globe deviated to the right, the excavation is deviated to the left and vice versa. In all of these cases, you continue the excavation opposite of the way the globe is viewed. The person who is experienced with this procedure will never go wrong in excavating a conduit which is aimed away from a specific shaft."

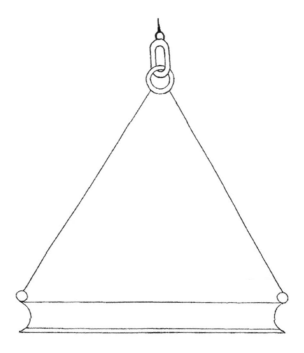

FIGURE 4.11   Brass tube level.

"On the other hand, to carry out the conduit excavation straight towards a certain shaft, a rope is placed on the ground and pulled straight from the shaft where the excavation begins to the shaft where the conduit should end. This rope should pass over the center of the mouth of both shafts. Then two strings, each with a hanging plumb bob, are lowered in the shaft where the poshteh digging is to begin. The strings should be about one zeraa shorter than the depth of the shaft. As the moghani continues digging, he views the strings from the end of the excavated section with one eye. Seeing the two strings as one, the excavation has been properly carried out. But, if he sees two strings, the excavation is out of line, in which case, he must realign the excavation towards the first string, namely the one which is closer to him. Excavating in dry soil, after digging a section along the two strings, he hangs the previously described tube level and views the strings through the tube to check if he sees the two strings as one (i.e., the strings are coinciding, author's note). Then he places a globe on the shaft wall along the strings, removes the strings and continues excavation looking at the globe through the tube. This way, he reaches the second shaft without any deviation or error."

4.6.1 EXCAVATION OF GHANAT

About types of soil Karaji notes (Karaji, pp. 34–35): "The best soil is that having uniform hardness and texture. A good soil is cohesive, fresh and has good odor. Adding sand and gravel to soil adds to its strength and suitability for carrying water. In such soils, the natural soil moisture avoids the water to dissolve soil."

"Any soil which contains natural moisture when left exposed to air loses its moisture, is no longer cohesive; and as soon as it becomes completely dry, it can get dissolved in water. However, ghanats and channels dug in dry soil would not be affected by flowing water, so long as their beds contain natural moisture. For this reason, when a ghanat or channel is dug in such soil, as soon as the excavated section is exposed to air, the soil loses its moisture; and if water flows through a dry section, it collapses. Therefore, if someone wants to dig a channel or ghanat in a soil which has natural moisture, he should allow water to flow through the dug section to keep it moist in order to avoid the soil losing its initial moisture. Evidently, the amount of water should be such that it would not hinder excavation. Many people and moghanis are not aware of this matter. That is they excavate the dry part of ghanat first and then begin digging the shafts in water bearing soil. Thus, the soil in the excavated dry section loses its natural moisture before the water from the test shaft (gamaneh) reaches there; and when the water enters the dry section, the walls of ghanat collapse. To avoid such collapse, the shafts in the dry and wet section should be dug concurrently so that the water flows continually in ghanat. This will help soil keep its natural moisture. If the ghanat soil is hard and solid, the solidity of soil helps maintain the excavated section. And, if the soil contains silt and gravel, the waterway will be more stable."

Karaji also notes: "Carrying out excavation in the water bearing reach of ghanat is easier than the dry section. The reason being that the water level in the ghanat reveals the ups and downs of its bed. In such a reach, moghani must follow a straight alignment and avoid curving right or left. Upon completion of ghanat excavation, the water depth should be the same from its beginning to the end and its crown should be straight."

The author had the opportunity to hold a round table discussion about the construction of ghanat with three experienced moghanis during a first national conference on ghanat which was held in Mashhad, Iran from June 27 to July 2, 1981 (Tir 1360 Iranian calendar). These moghanis had come from different parts of Iran, and among them, the one from Kerman (a city southeast of Iran), appeared to be the most knowledgeable. According to these moghanis, the construction of ghanat should preferably begin from its outlet (mazhar) to mother shaft especially if the ghanat has large discharge. It appears that the direction of constructing a ghanat depends largely on the type of soil (cohesive and hard or loose and soft) at the location of ghanat conduit. In hard rocky soil, digging can proceed in either direction. However, in cohesive soils such as clay, the conduit should be dug in downstream direction from the test shaft to keep the soil moist to avoid shrinking. Upstream of the trial shaft, digging has to be carried out in upstream direction in the water bearing strata. Digging in the opposite direction, water has no outlet and creates ponding. Occasionally, the water bearing stratum lies above the ghanat conduit upstream of the trial shaft. In such cases, the shafts digging has to be carried out both downward from the ground surface and upward from the ghanat conduit ceiling; since it is impossible to dig under water. As will be described later, this is the most difficult and hazardous task of ghanat construction.

In constructing a ghanat a number of crews may work concurrently. While one crew is digging the conduit from a shaft, another crew is digging the next shaft in the direction of the conduit.

The excavation in the conduit (poshteh) is carried out by moghani and his helper. Moghani, as indicated, uses a small pick, one side broad and the other side pointed, to dig the shafts and the poshteh. The excavated material is placed in a bucket similar to that described for the shaft excavation, pulled over the bottom of poshteh (conduit) until it reaches the bottom of the shaft. There, the bucket is connected to the hook at the end of a rope suspended from the windlass, is lifted to the ground and is emptied near the perimeter of the shaft.

Poshteh is commonly made large enough to allow a moghani to move in it in a bend down position. Therefore, ghanats which were made by shorter moghanis had smaller openings. Poshtehs are arch shaped, 75–80 cm wide and 130–160 cm high.[7]

In old days moghanis used animal fat (particularly from pigs and cows) burning lamps to lighten the ghanat conduit. Since the beginning of the twentieth century, kerosene and flash lights were substituted for animal fats. The lamps, apart from illuminating the work area, served as indicators of air quality in the conduit. Lights getting dim or off indicated poor air quality and insufficiency of oxygen. To correct this problem, the spacing between shafts were reduced which improved ventilation. The lamps also served as an indicator that the conduit is dug along a straight line. A lamp was placed on each side of the conduit to guide moghani to do his job. As the digging continued, moghani looked at each light from that side of the conduit; and if one of the lamps became partly visible, he changed the digging alignment slightly to correct the problem.

Digging the ghanat conduit was a very laborious, difficult, and also a dangerous job. Moghanis worked bent down day after day, month after month underground in low light and poor quality air. Because of the dust and dirt from digging, fumes from the lamps and dampness of the ground, the air was heavy and moist and endangers their lives. Also, in the water bearing sections of the conduit (poshteh), water dripped constantly on moghanis bodies.

Many Moghanis suffered from a general weakness and lung disease named silicosis later in their lives.

4.7 DIFFICULTIES OF GHANAT CONSTRUCTION

Ghanat construction may encounter a number of obstacles, including:

- Rock excavation
- Loose sandy soil
- Occurrence of excessive vapor and dampness
- Excavation below water bearing soil
- Excessive drip from conduit ceiling
- Loss of flow

4.7.1 EXCAVATION IN ROCK

Rocks are commonly large boulders which had fallen from mountains and covered with water borne silt and sediment over a long period of time (Wulff 1966). Moghanis overcame this obstacle with a significant amount of patience. If the rock was soft, they broke it with chisel and small sledgehammer. If the rock was at the center of the conduit (poshteh), they moved it to the side. If the rock was calcic, as Karaji indicates, Moghanis used heat to break down the rock. However, if the rock was solid and too large, they changed the poshteh alignment to bypass it. Moghanis had a lot of expertise in bypassing rocks in a poshteh. This was partly due to their accurate sense of location and partly due to their sensitive ears to hear the sound of pick beating in a nearby shaft.

To orient the poshteh's non-straight alignment on the ground surface, Karaji recommends the following procedure.

"Use a wooden or steel compass with perfectly flat and smooth inner and outer surfaces. Then pick a scale of proper length that is graduated exactly equal increments and enter the shaft from which the excavation of the conduit has begun. Tie a string to a nail driven at the center of the shaft and pull it straight to the first bend in the conduit until it touches a side of the conduit. Then drive a nail at that point, bend the string around the nail and pull it until you reach the next bend in the conduit. Drive a second nail at that point, loop the string around the nail in the same manner as the first bend. And continue this process until you reach the end of the conduit."

"The string should be long enough to cover the entire reach of the conduit. Obviously, the string makes an angle at each nail. Measure the angles by placing the hinge of the compass on each nail and open its handles along the string until the tips of the compass touch the string on either side of the nail. Then measure the distance between compass tips using the graduated scale (which is indicative of angle of the string, author) and record this distance. Likewise, measure the angles of string at each and every nail location using the graduated sale and write down your measurements. Then measure the length of the string from the center of the shaft to the first nail and also the distance from the first nail to the second nail and continue the measurements in the same manner from the last nail to the end of poshteh where you want to dig a shaft and take notes of the lengths of all sections. Karaji's sketch of the compass, measuring scale and measurement procedure in the conduit, is shown on Figure 4.12."

"Once measurements are finished, two strings with hanging plumb bobs are tied to a stick and lowered into the shaft. The stick is placed on the shaft and oriented such that when you look at the strings from the first nail in the conduit, they coincide on one another and appear as a single string. Then you exit the conduit, connect a string along the stick on the ground and extend it to a point such that the distance from that point to the center of the shaft is equal to the first measured length of the string in the conduit. Drive a nail on the ground at that location and bend the string around the nail to an angle exactly equal to that measured at the first nail inside the conduit. From that point, pull the string equal to the second measured length in the conduit and at that location bend the string at an angle equal to the second underground angle. Continue this process until the string on the ground surface follows the same shape as that of inside the conduit. The end of the string is the location where the shaft must be dug."

4.7.2 Loose Material

Another difficulty in ghanat construction is dealing with silt and sand and loose clay. To dig a shaft in loose soil, moghani uses open bottom boxes or baked clay hoops. As the digging continues, hoops or boxes are dropped in the shaft and pushed down to hold the loose soil from caving in. Excavating the conduit in such soils forms one of the most difficult parts of moghanis work, and the progress is very slow. It also poses the most danger to moghani's life. As there is always a likelihood of collapse of poshteh and burying the moghani. In such soils, moghani has to line the poshteh (ghanat conduit) with stone or arch sections (hoops). The arch sections (kaval or kûl) were traditionally made to baked clay 20 cm long and 1.25–1.4 m high and 75 cm wide.

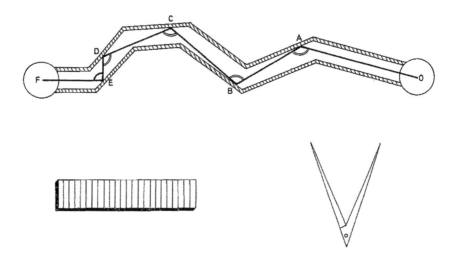

FIGURE 4.12   Compass, measuring scale, and measured polygon inside the conduit.

Figure 4.13 shows clay hoops. Karaji indicates that he had seen clay hoops in Isfahan (an old city in central Iran) with an opening so tight that a person could not enter them easily.

To protect the ghanat outlet from erosion in loose soil, the last five (5) or so meters of the ghanat is lined with stone. Also, as will be discussed later, ghanat conduit has to be lined where passing through loose sandy soil.

4.7.3 Excessive Vapor and Dampness

The following is Karaji's thoughts on the subject of vapor in the shafts and poshteh which inhibits excavation. (Karaji, pp. 57–58).

"In my opinion, three factors create excessive vapor and dampness in shafts and poshteh: First, the deepness of the shaft, second, the excessive length of the poshteh, and third, the occurrence of rotten soil. The closure of shaft mouths can also increase dampness in conduit. Rotten soil occurs where the soil has sulfuric substances or the soil contains traces of tar or the like which naturally create vapor and gas in the soil."

"There is vapor in any shaft or conduit in which the lamp turns off. The vapor is highest midday. The strongest lamp which can be lit in damp conduit is that which uses mumi or the fat of pig, cow or sheep. Second to these are olive oil or oil from oily grains. The use of tar is not recommended as it creates too much fumes."

"I have read in books by past generations that the best and strongest fuel for light in the vapor environment is olive oil, then mumy. When a moghani finds that the shaft dampness is temporary, he should place vinegar or watermelon (in the season) next to his work area in the shaft. If this does not lower the dampness, he should dig another shaft close to the one which has dampness and open a gallery from the new shaft to the old one. Alternatively, a long leather tube with the neck the same size or smaller than a spear cover should be used. One end of the tube is lowered to the bottom of the shaft and the other end connected to a strong smith blower at the top of the shaft. Working the blower while the moghani is digging will considerably reduce the dampness in the

FIGURE 4.13 Clay hoops.

shaft. Making the shaft square, which enlarges is opening, also reduces or eliminates the dampness. Good soils do not create vapor or dampness except at very large depths. Most often, the poor air originates from bad soil." "I have heard that in some of animal barns, sheep had died due to vapor from their own wastes."

"Deeps wells and long conduits do not have vapor in healthy soils."

4.7.4 EXCAVATION BELOW A WATER BEARING LAYER

Where a shaft hits a water bearing stratum before it reaches the required depth, the digging of the shaft may have to be carried out upside down. Specifically, the shaft is dug downward to the water level and upward from the conduit (see Figure 4.14). To carry out upward digging, moghani wears a steel helmet and a wet suit (a leather jacket impregnated with animal fat, in old days) and start digging the shaft from the conduit upward. Apart from its difficulty, this work is extremely dangerous as there is always the possibility of collapse of the shaft ceiling and burying the moghani. Moghanis refer to this work as devil digging (devil kan). Perhaps the word "devil," which implies ill action, has come from such digging.

4.7.5 EXCESSIVE SEEPAGE AND WATER DRIP

If large seepage is encountered along the conduit, moghani changes the conduit alignment slightly to avoid high flows. The alignment can be duplicated on the surface in the same manner as that of hitting a solid rock as described previously. The following are excerpts from Karaji (p. 59).

"Where, a large flow in the conduit or excessive water drip from the ghanat ceiling hinders excavation, moghani must wear a leather jacket made of treated calf skin and impregnated with melted cow fat. He also should carry a long edged hat made of the same leather to prevent water from flowing onto his face and his back. The rear edge

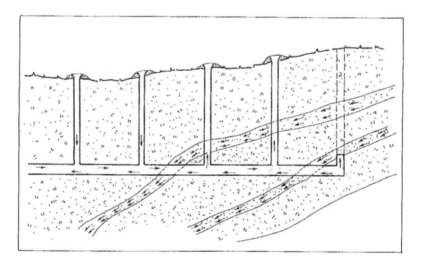

FIGURE 4.14   Excavation below water bearing stratum.

of the hat should be long enough to connect to his leather jacket and wrapped around it using a leather piece."

"If large amount of water seeps through the poshteh from all sides and the soil is non-cohesive, then baked clay hoops must be installed in the poshteh. The opening of hoops must be large enough to allow a person to enter them in a bent down position. Small holes must be provided around the hoops so that seepage routes are not blocked."[8]

4.7.6   LOSS OF FLOW

Along the conduit where the flow becomes small, moghani digs galleries on the sides of the ghanat conduit to intersect more water bearing strata. If necessary, more galleries are dug as the excavation of the conduit progresses. Some ghanats are 50 km long and contain up to 500 galleries. When the water table drops, the conduit is extended upstream toward the foothill to intercept a longer water bearing strata. Alternatively, galleries are dug radially from the mother well to increase the flow of a ghanat (Butler 1933). And, as the discharge capacity of ghanat is reduced, the galleries are extended radially.

It is worthy to note that Ranney well, namely a well with radial galleries, which was introduced by Leo Ranney, a Texan petroleum engineer in 1921, works based on the same principle as the galleries around the mother well, which was a common practice of ghanat construction in Persia several centuries earlier.

Where the ghanat discharge is lost excessively due to seepage in dry reaches of conduit, the ghanat bed is covered with impermeable materials such as bentonite clay or brick and mortar to diminish the loss. Alternatively, the said reach of ghanat can be lined with baked clay hoops. On this matter Karaji remarks:

"The lining would also be effective where the water has high salinity or hardness. If the salinity occurs within the wet reach, either the water has to be treated to lower its salinity or salt tolerant plants should be considered for cultivation," Karaji adds.

4.8   GHANAT SAFEGUARDING

In old days, moghanis were paying more attention than in recent times to safeguarding of shafts and ghanat conduits (poshteh). Albert Schindler (1877), a famous German traveler who visited Iran in the nineteenth century, reports of eight series of large ghanats near Damghan (in Semnan Province) which were constructed thousands of years ago. These ghanats were lined with bricks. The town residents at the time of his visit were removing the bricks to build their own houses. The reason for this care, similar to the installation of high walls, moats, and forts around towns, was to safeguard the ghanat from the destruction by invaders and enemies. Where the ghanat shafts and conduit were left unprotected, they could be readily destroyed by enemies in order to interrupt the life vessel, namely the water, of the town inhabitants and force them to submit.

Upon the completion of ghanat construction, the excavated soil is mounded around each shaft. The mounds avoid runoff during intense storms which carry silt and mud from entering the ghanat and contaminating its water. By avoiding high flows from entering shafts, the mounds also safeguard shafts and the ceiling and walls of poshteh from collapse which would result in blockage of poshteh and significant financial loss. The dirt mounds (known as karvar) are commonly 5 m wide and 1 m high. At deep shafts, these mounds were as wide as 10 m and as high as 2 m.

Where a ghanat is located along a gully, which concentrates the runoff during a heavy rainstorm, the mouths of shafts are covered with an arch brick structure with open top to prevent mud flow from entering the shaft while allowing the ghanat to aerate. A sustained rainfall, which occurred for three weeks during the spring of 1937 (Iranian calendar 1316) in Isfahan, resulted in collapse of nearly one-third of all the ghanats in that city, and some ghanats were totally destroyed (Beckett 1953). If the floodwater level is expected to rise above the mound around a shaft, the walls of the shaft should be lined with stone or brick. Alternatively, baked clay sections can be placed in the shaft and the poshteh. Where the ghanat is situated in an area where high winds are prevalent, the top of shafts should be covered in order to prevent air-borne dusts and sand to enter the ghanat. In such places the opening of the shaft is narrowed by placing rocks and the center opening is covered by a flat rock. In some cases, an arch ceiling using bricks is constructed on top of the shaft.

4.9   DELIVERY, INSPECTION, AND MAINTENANCE OF GHANAT

4.9.1   DELIVERY AND INSPECTION

The following sections are excerpts from Karaji's book (pp. 125–127):

> A newly constructed ghanat must be aligned along a straight line, have an even bed and have uniform water depth everywhere. The deeper water at any point along the ghanat conduit reflects a defectiveness and irregularity.
>
> In an improperly constructed ghanat, the water stands still at a location but moves fast at another. And where the soil is loose, that ghanat eventually collapses and is wasted. If it is not possible to dig a ghanat along a straight line, attempts must be made that the ghanat is constructed level, that the conduit ceiling be exactly or very nearly parallel to a horizontal line and the water depth is the same along the entire reach of the conduit.
>
> De-silting and cleaning are essential to ghanat longevity; and the qualified person, who wants to estimate the de-silting fees, must enter the ghanat beforehand and inspect its condition. And, depending whether the silt is hard or soft, negotiate an agreement with the moghani. He must estimate the fee carefully based on the length of ghanat; and after the moghani is finished de-silting, the person must enter the ghanat to inspect the entire work area and prepare a table of quantities.
>
> If the work has been performed satisfactorily in accordance with the agreement, he pays the moghani's fees. Otherwise, he adjusts payment based on shortcoming of moghani work. This is the good outcome of the prior agreement. However, the person who pays the moghani, based on measuring the extracted soil quantity from a ghanat, suffers a great loss. The moghani may have piled some dirt and placed a few buckets of the excavated soil from the ghanat on the pile to get it ready for inspection and the fee estimation by the owner of the ghanat or his foreman. Also, the moghani may have just picked up the dirt from the closely spaced shafts, avoiding desilting the farther reaches. If a moghani does not agree with the expert's assessment, his work would be idle. Likewise, it would be the work of someone who measures the mud from the shafts and claims that say this is from one-third of the conduit and its fee is this much or that is from one-fourth of the conduit and its fee is that much. A foreman who does not enter the shaft to inspect the moghani's work closely is considered to be a deceived and negligent person.

4.9.2 MAINTENANCE

Sediment consisting of silt and fine sand gradually cover the bottom of ghanat. Due to freeze-thaw action, most of the sediment originates from the walls of the shaft. The soil moisture at the walls of shafts freeze during winter; and as the ice melts in spring, it detaches the soil from the walls. The loose soil that falls to the bottom of the shafts is carried by flow through the poshteh and gradually settles. It is evident that the freeze-thaw action is stronger at the upper section of the shafts. Thus, over time, the shafts shape as a cone.

The sediment has to be removed regularly. Depending on the type of soil, this may be once a year, once in a few years or once in several years. To properly maintain a ghanat, the sediment deposit must be removed from the floor of the poshteh. This work which is performed by moghanis is referred to as silt removal or dredging (lair-oubi in Farsi). Over time, the ghanat section may become larger. In one of the ghanats in the City of Gonabad in Khorasan Province (northeast of Iran) the ghanat section is so large that a horseman can travel through it. On ghanat protection and maintenance Karaji writes (pp. 120–122):

> To maintain a ghanat, it must be de-silted and cleaned. More attention should be paid to the ghanat outlet (mazhar) where more silt accumulates and more algae grows on its bed. Another measure to maintain a ghanat is to cover the mouth of its shafts with brick and stone or construct an arch on the mouth using clay bricks. Also, during construction, the mouth of the shafts should be raised above ground using walls made of bricks, stone and clay mud. The excavated soil from the shaft and poshteh should be mounded behind the installed wall on the shaft mouth in order to prevent runoff from entering the shaft. The wall around the shaft should be sufficiently high and fully surround the shaft. This is the best measure to protect the ghanat from collapse. It also prolongs the life of ghanat and eliminates the ghanat owner from covering the shafts mouths.
>
> To remove the silt from a ghanat whose shafts are covered to protect it from runoff, first the covers should be removed and the shafts mouths kept open for a few days before entering the ghanat. Moghanis who work in damp shafts and poshteh to remove silt should eat light meals and avoid taking smelly foods which contain onion or garlic or similar odorous substances.
>
> It is to be noted that a majority of ghanat problems are due to collapse of shaft mouths. Another necessary measure for ghanat protection is to retain an inspector to enter the ghanat periodically to carefully inspect its inside so that if there is an excessive deposit of fallen soil, it is removed immediately. As a maintenance measure, ghanat should be inspected annually and cleaned as necessary. Where some of the shafts are located along a gully or concentrated runoff, their mouths should be raised using bricks and stones and a mound of excavated soil and mud should be built around them. The mud should be made using clay to prevent infiltration of water, or more preferably, the top of mound can be covered with stone.
>
> It is not advisable to build a ghanat along a flood route or in the midst of a valley unless the soil in ghanat is cohesive and hard. Building a ghanat in soft soil is a poor investment, unless the soil is moderately cohesive and the poshteh is lined with brick. If there is no choice other than to build a ghanat at the middle of a valley or flood route where water may rise over the mouth of their shafts and it may not be possible to cover the shafts as was indicated, the mouth of the shafts en route of runoff should be narrowed halfway with bricks and clay mud, as described as follows:

Dig on all sides of shaft and build brick or rock walls in the excavated area in such a way that when the sides are raised every one zeraa or so, they project inward three to four fingers. Then cover the top with a flat rock and cover it with clay mud up to the top of the shaft. An arch ceiling may be placed in lieu of the stone. The mud that is used in the wall must be made of a cohesive soil that has initial moisture.

Cleansing a ghanat may take a moghani from a few days to a number of weeks; however, a ghanat in Gonabad in Khorasan Province had not required any cleaning for over 100 years (Beckett 1953).

Estimates had indicated that the annual cost of ghanat maintenance is on the average 0.5% of its initial construction cost (Beckett, 1953; Cressey, 1958; Wulff 1968). The flow in ghanat normally fills less than one-fifth of the height of the poshteh. Under such condition, according to calculations by the author, the flow velocity varies from less than 1–1.5 km/hour (0.3–0.4 m/s) on the average. The discharge of ghanat typically range from 30 to 75 L/s.

However, during high flow (high water table), the water level in the upstream shaft rises considerably and the ghanat discharge may increase to 500 L/s or more. Such a discharge seldom occurs; however, if it occurs, it accelerates the erosion process.

4.10   GHANAT WATER RIGHTS/BUFFER

During Achaemenian and Sasanian Empires who ruled from 550 to 330 BC and AD 224 to 651, respectively, special attention was paid to ghanat. At that time there existed a perfect regulation on water distribution and farmlands, and water rights were recorded and kept at the tax department. After Arabs conquest of Persia in the mid seventh century AD, the ancient records of ghanat were ignored. Due to the lack of any knowledge and experience with ghanat, there was no consistent and meaningful ruling on ghanat ownership and its water rights during their rule which lasted through the latter half of the ninth century AD.

As was indicated in Chapter 3, after a terrible earthquake in the Town of Fergana, near Neyshabur (also reads Nishabur in English writings), A paragraph is added in AD 830 the people kept coming to Abdollah ibn Tahir (the son of Tahir ibn Husayn) from Tahirid Dynasty to intervene on their ghanats as they found no laws on ghanat ownership in clerical writing. To address the people's concern, Abdollah ibn Taher gathered all the clergymen from Khorasan Province and Iraq to compile a book on the ruling on ghanats which could be used to resolve a dispute. However, as Karaji indicated in his book, there were conflicting and improper rulings on ghanat ownership and its water rights.

To maintain the flow of ghanat it is important to enforce its water rights and its buffer. The extent of buffer depends whether there is another ghanat or a deep well near a ghanat. The buffer of a ghanat also depends on the rate of withdrawal of water from the deep well and its location relative to dry and wet reaches of a ghanat and the type of soil as well.

The withdrawal of water from a deep well can dry up a semi-deep well and that both of these can result in ghanat desolation. The buffer of a deep well is specified as 500 m in many parts of Iran and that of a ghanat is also taken as 500 m. However, the buffer of a ghanat should be significantly more than that of a deep well so that the withdrawal of water from deep and semi-deep wells, which results in lowering of the water table by several meters, would not reduce the flow of a ghanat.

As was previously indicated, the author attended the first Iranian seminar on ghanats in Mashhad (a city northeast of Iran) from June 26 to July 2, 1981 (month of Tir in 1360 Iranian calendar) and presented a paper on ghanats (Pazwash 1981). During this seminar, he held a roundtable discussion with three highly experienced moghannies from different provinces in Iran. According to these three experts and, in particular, the one from Kerman (a city in a southwestern province), the buffer of one ghanat from another one of the same level (termed hamrevesh, meaning that ghanat conduits are at the same or almost the same level), should be as follows:

1,000 m in clayey soil 1,200–1,500 m in sandy soil
1,500–2,000 m in coarse sand and fine gravel (they termed it as mouse teeth stones)
2,500–3,000 m in gravely and stony terrain.

Some other moghanis incorrectly opine the following: "The buffer of one ghanat from another of the same level should be 500 meter regardless of soil condition."

When the two ghanats are not at the same level, the above indicated buffers should be increased 500 m for each one meter of difference in their poshtehs elevations. Specifically, if poshteh of one ghanat is 2 m higher than another, the buffer should be increased by 1,000 m; and if there is a 4-m difference in elevation, the buffer should be 2,000 m more than the case of ghanats of same level.

Karaji's notes on ghanats' buffer were included in a previous chapter. Excerpts of his thoughts are iterated below:

> The ghanat buffer depends on the soil; that a ghanat may require no buffer if it is dug in a porous soil where its water originates from rain and snow and nearby rivers. Such is the case near the Diyala River south of Baghdad, where the water in the wells rise and drop with the increase or decrease in the river flow. The reason being that the water enters such ghanat from places near and far, from left and right and all around, especially if the ghanat is deep and water enters more from its side than its bottom.
>
> However, if a ghanat is dug in a flat broad plain and has no source other than rain and snow that falls on the perimeter mountains, it requires a buffer. Where a series of ghanats extend to the snowy mountains and there is no source of water along their route, the buffer for each ghanat is 500 Dheras on each side. Therefore, the distance between two parallel ghanats would be 1,000 Dheras.

If a ghanat is dug parallel to a mountain ridge, the entire distance between the ghanat and the mountain constitutes its buffer. However, the buffer away from the mountainside is minimal, as long as the bottom of the ghanat to be dug is not lower than the first ghanat.

The buffer of a ghanat from a deep well, as indicated, depends on the rate of discharge from the well, the depth of the well, the soil type and the location of the well relative to the dry or wet section of ghanat. A drop of water level as little as 5 or 10 cm at the location of water bearing reach of a ghanat may have significant adverse impact on the ghanat discharge. It is evident that the drawdown increases as the pumping prolongs. Therefore, calculating the drawdown should be based on the worst probable condition and sustained pumping.

4.11 GHANAT VERSUS WATER WELL

4.11.1 CONSTRUCTION COSTS

Prior to the invention of electricity and exploration of oil, the water used to be extracted from wells by hand pumps or buckets lifted by hand or a windlass. Since the twentieth century, deep water wells and electric or diesel pumps have become increasingly popular. Intuitively, for a given discharge, a single ghanat costs many times more than a deep well. In the 1954 water planning in Iran, for example, the cost of constructing a 4 km long ghanat, half of which would be in need of lining, was estimated at $15,000 and its annual maintenance at $370. The average flow of such a ghanat was assumed to be 30 L/s. In comparison, the initial cost of a 6" deep well with its pump and ancillaries was estimated at nearly $7,000; however, its annual maintenance cost was estimated at approximately $1,200.

A 35-km long ghanat in Ahmadabad (a village west of Tehran where the late Prime Minister Dr. Mosaddegh was put in house arrest) had furnished water to that village. The construction cost of this ghanat when completed in 1935 (Iranian calendar 1314)[9] was estimated at $8,000. This cost was twice as much as the worth of all dwellings and personal belongings of that small village at that time (Merritt-Hawkes 1935).

Also exemplified are two ghanats which were built in the 1940s and 1950s at the villages of Javadieh and Hojat Abad, south of the city of Kerman. Construction of the former began in 1941 and one crew of moghanis worked for seventeen years to bring water to the surface. In 1958, the owner hired a second team to work at night. Upon completion, the tunnel was 3 km long, its conduit bifurcated and had two mother wells 50 and 55 m deep. Because of loose sand, most of the conduit had to be lined with clay hoops and the cost of building this ghanat was $33,000 at that time. This was a monumental cost to build such a short ghanat. The cost of the latter ghanat, which had included 40 km long conduit and a 90 m deep mother well in Kerman, was $213,000 when completed in 1950 (English 1968). To construct the same ghanat in 1968 would have cost nearly five times as much.

At the 1978 prevailing labor fees, constructing a 20 km long ghanat would have cost approximately over $2 million. This is nearly equivalent to the overall cost of installing 10–15 water wells approximately 50 m deep including pumps and motors and other ancillaries. However, once a ghanat is constructed, it can deliver water day and night year after year with little or no supervision and minimal maintenance cost. It is astonishing that some ghanats, though over 1,000 years old, are still operational. The required periodic maintenance cost of ghanats is estimated on the average at 0.5% of their initial construction cost (Beckett 1953; Wuff 1966). Some ghanats have required little or no annual maintenance (Noel 1944) and a ghanat in Firuzabad in Khorasan Province, northeast of Iran, had required no desilting for over 100 years.

On the other hand, the average operating and maintenance cost of a deep well is estimated at 20%–25% of its initial cost. The above estimates indicate that with the annual cost of a deep well, a ghanat can be maintained for at least three consecutive years. Accounting for energy fees, the cost of supplying water by deep wells was approximately 3 ¢/m³ (8.5 ¢/CCF[10]) in 1968. At the same time, the cost of supplying 1 L/s of water by ghanat in Mashhad (the capital of Khorasan Province) was $170 annually. This amounts to 0.5 ¢/m³. The finished cost of ghanat water varies based on soil condition and the extent of maintenance. However, the average cost of water was about 1.0 ¢/m³ which was one-fourth to one-third of that of deep wells.

4.11.2   ENERGY COST OF WELLS

Unlike ghanats, deep wells require a continual supply of energy, whether electricity or fuel to run their motors. Calculations for the required electrical and fuel needs are presented in the following paragraphs. The calculations are based on 750 m³/s, which represents the ghanats discharge prior to 1962, an average water well depth of 50 m and 245 days (8 months) of growing season, which is nearly equal to the average cultivation period in Iran. The energy consumption for electrical and fuel powered motors are as follows:

A. Electrical Pumps:

Lifting 1 m³ of water by 1 m requires: $1,000 \text{ kg} \times 9.81 \text{ N} \times 1 \text{ m} = 9810 \text{ W} = 9.81 \text{ kW}$

The actual energy requirement for electric pumps with an overall efficiency of 65% would be:

$$9.81/0.65 = 15.1 \text{ Kw}$$

Therefore, supplying 750 m³/s against 50 m total head as shown below requires 3.33 billion kWh of energy annually.

$$750 \times 50 \times 24\text{hrs} \times 245 \times 15.1\text{kW} = 3.33 \times 10^9 \text{ kWh}$$

In the absence of reliable data in Iran, the cost of energy is based on the US rates. At the current (2024) average gross electricity rate of 16¢/kWh in the US (accounting for the service charge), the annual cost of energy alone would be over $530 million.

B. Fuel Pumps

Fuel requirements for pumping with internal combustion engines are calculated based on the calorific value of fuel. Since one[11] horsepower (hp) equals 75 kg m/s (735.5 watts) or 175.7 cal/s, lifting water at a rate of 1 m³/s by 1 m requires:

$$1000 \times 1\text{kgm} / 75 = 13.33\text{hp}$$

or

$$13.33 \text{x} 157.7 = 2,343 \text{ cal/s}$$

This energy must be divided by the efficiency values for pump to obtain the required energy input for the engine.

Gasoline motor pumps use 1.7 cm³ of gasoline every second to lift 1 m³ of water by 1 m. For diesel pumps, the need is estimated at 1.2 cm³ [12] (Bouwer 1978). Therefore, lifting 750 m³ of water by 50 m, as shown below, would require nearly 10 million barrels of oil annually.

$$750 \times 50 \times 3600 \times 24 \times 245 \times 1.7 / 1000 = 1.35 \times 10^9 \ \text{L}$$

$$1.35 \times 10^9 / (3.78 \times 42) = 8.5 \times 10^6 \ \text{barrels} = 357 \ \text{million gallons}$$

where 42 and 3.78 represent the number of US gallons per barrel of oil and liters per US gallon, respectively.

Using diesel pumps, the fuel consumption would be:

$$357 \times 1.2/1.7 = 252 \ \text{million gallons}$$

At current (2024) average oil prices which is approximately $3.44 per gallon for gasoline and $3.71 per gallon for diesel, the gasoline and diesel fuels alone would cost $1226 and $933 million, respectively.

4.11.3   ENVIRONMENTAL IMPACTS OF WELLS

Apart from cost, the environmental impacts of ghanats and deep wells deserve consideration. Since flow in ghanats occurs naturally due to gravity, there is no adverse impact on ground water resources. However, since the discharge from deep wells is controlled by man, overdraft occurs and both the quantity and quality of groundwater is impacted. Another, as important if not more important, advantage of ghanats is that they require no power and no interruption in service, whereas deep wells are subject to interruption due to failure of pumps and motors; and in remote rural areas, may take days to be put back into operation.

In addition, ghanats have the following environmental bonuses:

- No air pollution.
- Far smaller loss of water compared to channels.
- Keep water fresh and free of pollutants such as salt; and especially, wind-blown dust which is a major airborne pollution in barren lands.
- Provide a niche for the vectors that transmit water borne diseases that so seriously affect the population of areas irrigated by modern technological means (Goldsmith and Hildyard 1984).
- Maintain water at almost constant temperature; and as such can also serve as a thermal power in cooling buildings in summer and providing some heat in winter. This matter was exemplified in Chapter 3.

In short, ghanats are the most ecological and environmentally friendly and sustainable means of extracting water.

A concern expressed by some agriculture specialists is that ghanats discharge exceeds the demands during winter months and the water goes to waste. This concern, however, has no justification because the groundwater flows due to gravity until it reaches a flat plateau and it becomes stagnant. Because of mixing with minerals on the way, the water there is saline or hard and unsuitable for use anyway.

Moreover, the ghanat water in part, had been used for many centuries to:

- Store water in underground cisterns.
- Make ice in natural underground freezers (ice house, called "yakhchal" in Iran). See Figure 5.6 in the next chapter.

In addition, ghanat water can be used to:

- Flood fields and agricultural plots in winter. This does not only promote infiltration and groundwater recharge, but also creates thick frost which helps destroy fungus. It also avoids premature greening and tree blossoming in spring.
- Fill natural depressions and man-made ponds during winter to increase supply of irrigation water in summer.

NOTES

1 The word qanat appears widely in the English language. However, the word is pronounced as "ghanat" in Iran.
2 In the English language, this word appears as moqanni; however, it is pronounced "moghani," in Iran.
3 Karaji's book also contains more detailed information about the type of soil and vegetation and color of rocks and mountains as indicators of groundwater occurrence. This information was presented in Chapter 2 entitled "Excerpts of Karaji's Book."
4 It appears that such soils contain phosphorous.
5 See note 15 in Chapter 2.
6 Tabatabaee was Mayor of Tehran and the Prime Minister of Iran from February to May 1921, during Ahmad Shah, the last Shah of Qajar Dynasty.
7 The water supply tunnel of Montville, New Jersey, which was constructed around the 1890s, was 1.7–1.8 m high. The author walked this rocky tunnel with a head lamp in 1986 to find a solution to eliminate the contaminated groundwater from entering the town water supply system. The chemicals from the aircraft instrumentation at the Moratto Scientific Plant that had infiltrated into the ground was seeping into the water tunnel. Upon considering various options, the author recommended placing a 12" PVC water line inside the tunnel. This solution was not only most effective, but it was also the least costly and was implemented in a short time.
8 The same idea of holes in hoops is employed in the construction of retaining walls where weep holes are provided to allow free flow of seepage in order to avoid buildup of water pressure behind the wall.
9 Iranian calendar was changed from Lunar (AH) to Solar Hijri (SH) on February 21, 1911, and was legally adopted on March 31, 1925 under Reza Shah Pahlavi. This calendar, like Gregorian, has 365.2422 days.
10 CCF = 100 cubic feet.
11 The efficiencies indicted in this reference are too conservative. Efficiencies of 35% for gasoline pumps and 45% for diesel pumps are more realistic.
12 Based on 35% and 45% efficiencies, the cost of gasoline and diesel fuels would be $570 and $430 million, respectively.

5 Irrigation and Water Storage in Persia

5.1 INTRODUCTION

For survival man has searched for water as early as mankind itself. Where there existed surface water, such as lakes and rivers, man settled and built communities. Many of the major ancient cities in the world, such as Damascus in present Syria, Beijing in China, Kolkata in India, Alexandria in Egypt, and Babylon and Ctesiphon (Tisphon) in Persia (which is now Madaen in present Iraq), were built along major rivers. In dry parts of the world, man searched for another source of water, namely the water below the ground surface. This search most likely began with digging a hole at or near a spring to explore its source and was gradually developed to well construction in China and Persia.

The well digging over time was expanded to include a horizontal gallery to increase its capacity. Eventually, man learned to connect wells by horizontal galleries and bring groundwater from hillsides to playa skirts. Thus, the technology of qanat (also called kariz and ghanat in Iran and qanat in Arabic) was invented. While the Chinese advanced the water well technology, Persians professed in the art of qanat construction.

In addition to qanats and irrigation channels, Persians built cisterns, both large and small, to store water and also built dams and weirs to control rivers for irrigation. The origin and construction of qanats were discussed in previous chapters; only a brief description of qanat advantages together with water cisterns, ancient irrigation canals, weirs and dams are presented in this chapter. Also described in this chapter are construction materials that were used in ancient Persia.

5.2 IMPORTANCE OF IRRIGATION IN PERSIA

In the Avesta, the sacred book of ancient Zoroastrian Persians, irrigation (āb yāri) was a good deed in the eyes of Ahūra Mazda (the God of Zoroastrians). Wastelands and deserts were considered as haunted by Ahriman (evil) and its demons. The Achaemenid Kings (550–330 BC) exempted from land tax for five generations to all who made land cultivable through the construction of an irrigation system (Von der Osten 1956).

Construction and maintenance of ghanat and irrigation channels undoubtedly had required implementing cooperative rules and regulations that had been developed over time. Depending on geographic and economic variation, these rules had been somewhat different from place to place in the Persian land. However, the rules were invariably based on the water availability and its equity of use.

DOI: 10.1201/9781032659930-5

During Sasanian (Sasanids, misspelled Sassanids in most English writings) Empire (AD 224–651) and just before the Arabs' invasions (which lasted from AD 633–651), there existed many laws, regulations and customs that governed the construction of irrigation and water supply systems and their maintenance and equitable distribution of the available water. The traditional system was so highly valued that the Soviet Union paid particular attention to it for planning irrigation works of Kazakhstan Republic, formerly a part of Persian land.

As was indicated in a previous chapter, these regulations were totally ignored during the Arab's rule in Iran. Instead, religious laws (Sharia) were established, which as indicated, were somewhat contradictory. To address conflicts and complaints by farmers and qanat owners, as noted in Chapter 3, Abdollah ibn Tahir (the son of Tahir ibn Husayn) who ruled Khorasan under Abbasid Caliphs, gathered a group of clergymen from Khorasan and Baghdad to put together rules for qanat maintenance. The rules, titled Al-Qanun val Nahr, were partly adopted from Sasanian. It is evident that since there existed no qanat or canal in the desert of Arabia, there were no Islamic rules for same.

5.3   GHANATS

In deserts and broad playas where the ground surface is flat, groundwater becomes stagnant and remains in contact with soil for long periods of time. Consequently, the water dissolves the salt and minerals present in the soil and may become saline or hard and unsuitable for use. At foothills where the water table is steep, the groundwater is continually moving and water is fresh. For this reason, all qanats in Persia and elsewhere are constructed originating at foothills so that they collect freshwater and convey it by gravity to playas. Since the water in a qanat moves from foothill to point of discharge within hours, the water remains fresh. Thus, in addition to their economic and environmental benefits that were discussed in a previous chapter, qanats also avoid contamination of water by salt and minerals that are present in the soil. Apart from supplying water for irrigation and domestic needs, ghanats were also used for cooling rooms and running water mills. Through ghanats, Persians supplied water for irrigation and turned deserts into fertile lands and food centers

5.4   IRRIGATION CHANNELS AND CONDUITS

In the old days Persians paid more attention to construction of irrigation channels than in recent times. Karaji's description of constructing open channels and closed conduits are exemplified in this chapter.

5.4.1   OPEN CHANNELS

On constructing a channel in sand and seeping soil, Karaji notes (Karaji, pp. 62–63):

> Where a channel (called jub or nahr in Iran) traverses through loose soil, the channel should be covered with large bricks and gray lime (called sarooj in Iran). Gray lime is a lime which is mixed with a smaller amount of ash from lime baking furnace. Before mixing, the gray lime should be made fine using a steel tamper. The bricks should be

placed perfectly flat on the bottom. The channel sides should also be covered with bricks and lime mortar. Alternatively, depending on the amount of flow, the soil may be removed up to a depth of one zara or less from the bottom of the channel and replaced by clay mud. Then the mud layer should be compacted to the original level. Likewise, the sides should be covered with clay mud in a curved shape up to the water level.

Using a clay mud which has initial moisture to cover the sides will add to the stability of the lined channel provided that the sides are watered so that the soil maintains its initial moisture. Adding fine gravel and sand to clay soil before compaction will increase the stability of the lined channel. Our predecessors have indicated: leave cattle to walk over loose soil to get soil compacted under their feet (referring to the need for compaction, author).[1] Also, if the clay soil that has initial moisture, is mixed with equal amount of dry lime and the same amount of sand, placed at the bottom of the channel and is compacted using a steel tamper and the water is allowed to flow through the channel, it will become as hard as a rock over time. Sometimes, the bottom becomes so hard that it makes it difficult for the moghani to excavate the soil. Also, the soft soil may be covered with flat rocks and the gaps are filled with a mixture of clay, sand and lime.

To eliminate seepage in a drainage channel in Dezful, the sides and bottom of the channel were lined with asphalt. In the summer of 1963, the author evidenced that this liner, which was designed by a foreign consulting firm, had become loose under sun rays and collapsed from the banks. The designer neglected to realize that the air temperature in a summer- day risers to 55°C (131 F°) and the asphalt becomes even hotter than that air under the sun.

5.4.2 CLOSED CONDUITS

On this matter Karaji writes (Karaji, pp. 60–61):

We then say that tanbusheh (short clay conduits), in old days are installed along a water route for two reasons: first to eliminate high seepage from an open channel, and second to avoid water contamination where the channel traverses through a developed area. The first subject to be discussed herein is the shape of the conduit and how it is made.

The conduit (tanbusha) should be made such that one of its ends is larger than the other so that the narrower end can fit two finger widths inside the larger end. The conduit should be four times as long as its larger diameter; however, it can be longer and stronger if the soil has high cohesion. The wall thickness of the narrower end should be less than the other end. The conduit should be made using fresh soil having no gravel and should be fully baked. A pure soil whose sand and gravel are removed by washing makes stronger conduit.

On the installation of these conduits Karaji writes:

The conduit alignment should be excavated in such a manner that there are no ups and downs when measured with a level and string. Also, the location of the end of a tanbusheh should be slightly lower than its entrance so that the water can flow naturally. The first tanbusheh is placed in the excavated channel in such a way that water enters at its larger mouth and exits the narrower one. The narrower mouth should be covered with lime mortar (prepared as will be described later) within two finger widths before placing it in the larger mouth of the next section. The joint should then be covered with the same mortar. Also, a vent should be placed at every 100 zara of conduit to prevent buildup of air and vapor which may cause fracture of a tanbusheh.

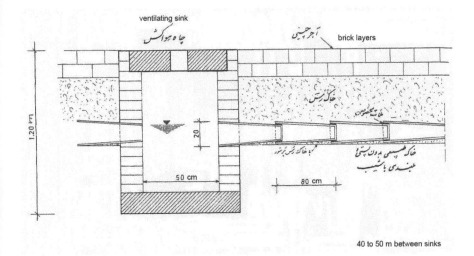

FIGURE 5.1 A sketch of baked clay conduits (tanbusheh) in Persia.

Figure 5.1 shows a sketch of tanbushehs (short clay conduits) alignment and air vent.

"When the installation of conduit is completed, it should be left alone for at least three days before water is allowed to slowly enter it. Rubbing inside the tanbusheh with oil or melted fat before installation will add to its durability and capacity in carrying water. Upon installing the conduit, the sides should be covered with clay so that no opening is left at its sides and bottom."

In 2014, scientists performing rescue archeology at Farash historical site at the Seimareh Dam Reservoir uncovered remnants of a water supply conduit. This system comprises of a small pool and baked clay conduit pieces, each piece measuring one-meter long. The system was located at the eastern shore of the reservoir on the border between Ilam and Lorestan Provinces. Before filling the reservoir, part of the system was covered in order to be saved for future archaeological excavations when the dam is decommissioned. Iranian archeologists claimed it to be a water conduit from the late fourth millennium BC. When proven, this would turn the irrigation and water conduit technology some 2,000 years back in time, over 5,000 years ago (Payvand Iran News). Also, it has been reported that the first irrigation project was carried out at Ghogha Mish Hill (between Dezful and Shushtar) in Khuzestan and that this irrigation system was destroyed 6000 BC.

5.4.3 Preparation of Lime Mortar for Conduits

On this matter, Karaji writes (Karaji, pp. 61–62):

> Where the use of mortar is necessary, the lime rock which is mildly baked is sprinkled with water to break up. Then the sprinkled lime is powdered and filtered with a fine sieve and to every twelve kilograms of lime, one kilogram of olive oil or another vegetable oil, but preferably olive oil or milk, is added and mixed together. Then the lime mixture is placed in a pestle and tapped with a wooden handle and more oil is added as needed. This mortar should be used immediately after preparation so that it is not

dried and wasted. Adding eggs[2] to lime increases its durability; likewise, using more oil adds to the lime's strength.

One of my predecessors had said that: Adding a small amount of vinegar to the water to be sprinkled over the lime increases the lime's durability; another person had said that: A paste made of grinded iron rust mixed with egg whites is very good substance for plugging cracks and holes in pools, basins and ponds. However, rather than sprinkling the lime with water, if it is tampered and mixed with oil to make a paste and used immediately, it will be stronger and more suitable in filling seams and gaps of containers and the like.

5.5   WATER CONSERVING IRRIGATION METHODS

In Yazd, a genuine ultra-conversing method of irrigation had been devised. This method, which Pazwash (2016) has referred to as jar irrigation, basically involved using a clay jar with narrow neck. The clay jar was buried next to a plant and filled with water as needed (on the average, once every two weeks, author's note). Seepage of water through the jar kept the soil at a constant moisture at or somewhat near its field capacity, namely the amount of water a soil can hold against the force of gravity. Thus, no excess water was lost due to percolation and there was no evaporation from a wet ground. This method of irrigation also eliminated any loss of water due to infiltration that would otherwise occur in the conveyance channels and ditches in furrow irrigation. Jar irrigation used to be employed for watering small vines and bushy plants and melons.

In addition to jar irrigation, Persians planted seeds of melons and the like next to the roots of phreatophyte (deep-rooted) plants such as Khar-Shotor. The deep-rooted plant extracted groundwater deep underground and fed the roots of the plants intermingled with its roots near the ground surface.

5.5.1   Irrigation Measuring Devices

It is known that the Persians were using water clocks in 323 BC to ensure a just and exact distribution of water from qanats to their stakeholders for irrigation. In fact, the use of water clocks in Gonabad qanat and Zibad qanat dates back to 500 BC. Later, they were using water clocks to determine the exact pre-Islamic Persian holidays[3] such as Norooz (spelled Nowruz in English) and Yalda (the first day of winter when the day begins to be longer than night). The water clock or Fenjan (cup) was the most accurate and commonly employed timekeeping device for calculating the amount of time a farmer must take water from Gonabad qanat until it was replaced by a more accurate current clocks. Figure 5.2 shows an ancient water clock. Water clocks consisted of a vase with a hole at its bottom. The vase was filled with water and allowed to drain through the hole. The time it took the vase to empty was used as a measure a farmer could take water from a qanat. The number of times that the vase would be filled and emptied was related to the size of a farm.

The size of the vase and its hole differed from town to town.

In addition to water clocks, water measurements were made volumetrically. A typical measure, called "sang" varied from place to place; for example, it was 20 L/s in Tehran, whereas 16 L/s in Shiraz. In 1941 (1320 SH),[4] when the issue of

FIGURE 5.2   Ancient Persian water clock (Public domain/Wikicommons).

irrigation water was revisited, engineers adopted 13.3 L/s as a unit for sang. This implies that $1\,m^3/s$ amounts to seventy-five sangs. Sometimes a sang is rounded to 15 L/s.

5.6   CISTERNS (AB-ANBARS)

To cope with the absence of perennial streams and scarce precipitation, ancient Persians had learned to store rainwater or stormwater runoff or qanat water in cisterns. Until recently, the use of cisterns was a common practice of water storage at homes and villages. As a child, I remember that our old home, which was built around 1910s, had a cistern. A conduit made of vitrified clay (tanbusheh) connected a channel downstream of a qanat to our home. The cistern was built underground next to the kitchen and was filled a few times a year. This water was used for all domestic needs other than drinking. Upon filling, several large pieces of rock salt were dropped in the cistern to keep the water fresh. There was also a pool at our home which was filled as necessary by qanat water and supplied irrigation needs. Many homes in Tehran had cisterns (ab-anbar, meaning water storage) until the municipal water supply was constructed in 1953. Ab-anbars are still used in many rural areas in Iran.

To meet the needs of caravans and travelers, large ab-anbars were constructed along roads in deserts. Large ab-anbars were also built in villages, castles and royal palaces. There was an underground vault under Apadana Palace in Persepolis which some incorrectly perceived it as a jail for enemies or burial place. It is likely that the vault served as a cistern for harvesting the roof rain.

Ab-anbars generally consisted of a cylindrical shell underground and a cone shaped arch structure above it. The storage volume of public ab-anbars generally ranged from 300 to $3,000\,m^3$, however occasionally much larger. A 20 m diameter, 10 m deep ab-anbar provides over $3,100\,m^3$ of storage. Large ab-anbars were made of two chambers underground with two domes above ground. Figure 5.3 shows a large beautiful ab-anbar with four air towers in Yazd.

At places structures similar to ab-anbars were used for ice storage. To make ice, water was entered on a flat surface surrounded by tall walls on the south, east and west

FIGURE 5.3   A large ab-anbar in Yazd (photo by Shahab Etemad).

sides to hide ice from sun rays. Every day in winter, water was sprayed on ice to cover it at shallow depth to freeze overnight. When the ice thickness would reach 30–40 cm, the ice was broken in large chunks and stored in ice cisterns (yakh chals) for use in summer. Before the invention of electricity, this was the only way to make and store ice; it was also an environmental-friendly means of ice making. Figure 5.4 shows an old yakh chal, made of unbaked, sun-dried clay bricks (called khasht in Iran) in Kashan, Iran.

FIGURE 5.4   An old yakh chal, made of sun-dried clay bricks in Kashan.

5.8 CANALS

5.8.1 GENERAL

Irrigation channels were vital to Mesopotamia for the land between the rivers (called mian rudans, meaning between rivers) during Sasanians. Flooding was serious in Mesopotamia and the Tigris and Euphrates Rivers carried a great quantity of silt. Laws in Mesopotamia not only required farmers to keep their canals and basins in repair but also required everyone to help with hoes and shovels in times of flood or in digging new canals or repairing the old ones. Some canals were used for 1,000 years before they were abandoned. The following sections briefly describe some of the most important ancient canals, whether used for irrigation or navigation.

5.8.2 SHUSHTAR HYDRAULIC SYSTEM

Shushtar historical hydraulic complex is a vast ensemble of flow control structures and irrigation canals. It includes an interconnected set of bridge-weirs, canals, tunnels, water mills and water cascades, among them were Daryoon and Gargar Canals, Salasel Castle and Mizan Weir. This complex is situated at the foot of the Zagros Mountains near the City of Dezful and approximately 90 km north of Ahvaz which since Sasanid (Sasanian) Empire has been the Capital of Khuzestan Province.

The primary construction of this complex, which included Daryoon canal ad Salasel Castle, is attributed to Darius the Great (Daruish Bozorg) of Achaemenid Empire five centuries BC. Daryoon canal, which branches from the Karun River, originates from two tunnels, each 3–4.5 m wide which carry water from the Karun River, extend under Salasel Castle, and merge 100 m or so past the castle. Figure 5.5 shows

FIGURE 5.5 Daryoon weir bridge over the canal; Salasel Castle at background.

the weir bridge over Daryoon Canal. This canal, which is still partly operational, has provided water to irrigate Minab Plain. The Salasel Castle was the operation center of the irrigation system in Shushtar. The castle had included a tower for water level measurement, numerous yards, stables, bathhouses, gardens, an armory, a kitchen and a moat. It was also a control center for Daryoon Canal, which as was indicated, was another man-made water course branched from the Karun River and was constructed during Darius the Great. Today, the castle is ruined; only the underground rooms and Daryoon Tunnels have remained.

Other structures including Gargar canal and Mizan weir (called Band-e Mizan, in Iran)are known to have been constructed during Ardeshir I, the founder of Sasanian Empire in the third century AD. This complex was constructed using granite rocks and a plaster of lime mortar.

Karun, which is the largest and the only navigable river in Iran, is divided into two branches, Gargar canal and the Shatit River, before entering Shushtar and form this city as an island. To construct Gargar canal, the flow of the Karun River was bypassed through a man-made diversion channel. During the diversion, which took nearly three years, the bed of the Karun River was dried out, raised and paved using large stone blocks by a few meters in order to allow water to enter Gargar canal and irrigation channels. Within the same period, Band-e Kaisar, also called Pol-e Shadorvan or Shushtar weir bridge, was built over the dried-up river bed. Also, a weir bridge named Mizan Weir Bridge (Band Mizan) was constructed on the Karun River just upstream of the confluence of Gargar Canal and Shatit River, which is the main branch of the Karun River. Mizan Weir Bridge is an archeological and architectural masterpiece. It is so constructed as to distribute the Karun water between Gargar and Shatit; and when the water in one of these exceeds a certain amount, divert it to the other.

The Shushtar weir bridge (Pol-e Shadorvan), which was constructed during Sasanid Empire, is located shortly downstream of Daryoon Canal. It carried across the important road between Pasargade (the capital of Achaemenid) and the Sasanid capital, Ctesiphon. This bridge was over 500 ft long and included 40 spans. Figure 5.6 shows a picture of the upstream face of two spans of this historic bridge, taken over 50 years ago.

Past Mizan Weir Bridge, the Gargar canal first crosses the rocky bank adjacent to the city of Shushtar and then the water is conveyed through a series of tunnels that run mills and supply water to the city of Shushtar. The waters from the mills fall from a cliff onto the basin in the form of a water cascade. Figure 5.7 shows an spectacular and fascinating view of some of these water falls. The Gargar canal then enters the plain south of Shushtar at the foot of the mountains, where for almost 1,800 years its water has contributed to planting orchards and fertile farmland called Minab (literally "The Paradise"). After nearly 50 km, Gargar canal (also known as Gargar River) and the Shatit River merge.

The Achaemenids, and more so the Sasanids, used genuine technologies to share the Karun water among different areas in Shushtar and some of the nearby cities. Through these, a desert like land was made habitable and vast areas became irrigable. In Shushtar alone, over 40,000 ha of land in Minab was thus irrigated. Alas, hectares and hectares of wetlands and farm lands were taken for oil exploration in some decades ago; and as a result, dust storms were greatly aggravated in Khuzestan.

FIGURE 5.6 Two spans of Shushtar Weir Bridge over the Karun River, built during the third century AD.

FIGURE 5.7 Canal in Shushtar and outfalls from water tunnels Photo by Shahab, Etemad, taken in 2016.

The Shushtar hydraulic complex is very unique and draws many visitors and archaeologists from around the world to the area. The International Scientific Committee on Archaeological Heritage Management (ICOMOS) has recognized the outstanding universal value of Shushtar complex both in terms of its integrity and authenticity (Shushtar Hydraulic System, Iran No. 1315). UNESCO sited Shustar Historical Hydraulic System as "a masterpiece of creative genius" and describes it as follows:

> Shustar, Historical Hydraulic System, a masterpiece of creative genius, can be traced back to Darius the Great in the fifth century BC. It involved the creation of two main diversion canals on the river Kârun one of which, Gargar canal, is still in use providing water to the city of Shushtar via a series of tunnels that supply water to mills. It forms a spectacular cliff from which water cascades into a downstream basin. It then enters the plain situated south of the city where it has enabled the planting of orchards and farming over an area of 40,000 ha known as Minab (Paradise). The property has an ensemble of remarkable sites including the Salâsel Castel, the operation center of the entire hydraulic system, the tower where the water level is measured, dams, bridges, basins and mills. It bears witness to the know-how of the Elamites and Mesopotamians as well as more recent Nabatean expertise and Roman building influence.

Jane Dieulafoy, the famous French archeologist, in her travel has referred to this as the largest industrial complex.

5.8.3 ANCIENT SUEZ CANAL

The ancient Suez Canal or Necho's Canal, which linked the Nile River to the Red Sea, was a navigation, rather than an irrigation canal. According to Herodotus (a highly respected ancient Greek historian) the canal was constructed under the Persian King Darius the Great five centuries BC who ruled Egypt at the time. Egypt was conquered half a century earlier by Cambyses II (Kambiz in Persian), the second Achaemenid King and son of Cyrus the Great. While some later authors, like Aristotle and Pliny, claimed that Darius did not complete the work. Daruis's Suez Inscription indicates otherwise. The inscription reads:

> I am a Persian; setting out from Persia, I conquered Egypt. I ordered to dig this canal from the river that is called Nile and flows in Egypt, to the sea that begins in Persia. Therefore, when this canal had been dug as I had ordered, ships went from Egypt through this canal to Persia, as I had intended.

The ancient canal allowed passage from the Mediterranean Sea to a Nile Delta tributary, then through the canal to the Bitter Lakes, then south to the Gulf of Suez. The current Suez Canal which was dug and opened by the French in 1869, goes directly from the Mediterranean north to south to the Gulf of Suez for 120 miles.

5.8.4 XERXES CANAL

Another strategic canal worth mentioning is a 2-km canal built through the base of Mount Athos Peninsula in northern Greece by King Xerxes (pronounced Khashayar in Iran) of Achaemenid Empire in the fifth century BC. This canal is conceivably one of the biggest engineering works of the time and one of the few monuments left by the Persian Empire in Europe. It was built to conquer Greece, Herodotus notes. Ever since, historians have debated whether the Xerxes Canal was really dug from coast to coast. Some have even doubted its existence reasoning that rocky plateau made its construction an impossible task at that time.

However, scientists from Britain and Greece have come up with conclusive evidence that the canal was indeed built. Through excavation, the scientists have drawn a map which show the canal's dimensions and course. The findings confirm the descriptions by Herodotus, which some scholars have questioned with skepticism.

Spanning about 30 m on the surface, the canal was sufficiently wide for two war galleys to pass. The canal sides slope downward to a depth of approximately 14 m forming a width of roughly 15 m at the bottom (Y. Bhattacharjee, New York Times, November 13, 2001). This implies a side slope of 1 m horizontal to 1.9 m vertical (author).

5.9   WEIRS, BRIDGES, AND DAMS

5.9.1   AN OVERVIEW

Most of the rivers (ruds) in central and south Iran are seasonal. As such, dams and weirs were constructed on streams throughout history in Persia. These structures store snowmelt and the surplus of water in spring, when rain was more plentiful than other seasons, and take it directly to field to supply streams (stream is called jub or juy in Iran). Some of these dams are still functional though at a reduced capacity due to siltation over time. One of the functional dams, as will be described later, is Band-e Amir in Shiraz which was built about AD 960 by AmirAzad-al Daula (Dawla) Daylami (Buyid dynasty, Persian Al-e Buye). It is known that the civilization in ancient Persia was centered around the southern part of Fars Province near Persepolis, which was the capital of Achaemenids, and the southwestern part of Khuzestan Province, which contains many rivers and streams. In those areas there were many old weir bridges (pol-band) which date back to nearly five centuries BC and some of which still exist. These weirs had a significant impact on the irrigation and reclamation in this area.

There are three large rivers and a number of small streams in Khuzestan. Among the large rivers, the Karun River (the most effluent and the only navigable river in Iran) has the largest discharge and Kor and Abdez (or Dez River) have intermediate flows. In the old days, Khuzestan was a production center of grains and orchards. Although some of the streams and old canals have dried up because of construction of new large dams since the 1950s, the remnants of some of the old weir bridges (and dams) still exist. Many of the old structures on rivers were constructed to function both as a weir and a bridge. They served as a stream crossing and also raised the water in the river. Figure 5.8 depicts typical plan and profile of one such dual purpose bridge-weir (pol-band) structure going back to Sasanian Empire, third century AD. The upper arches in this structure were intended to provide additional flow capacity and to reduce the quantity of rocks/bricks above piers. These arches also reduced the load on the lower arches still functions perfectly without any major repair for over three and a half centuries.

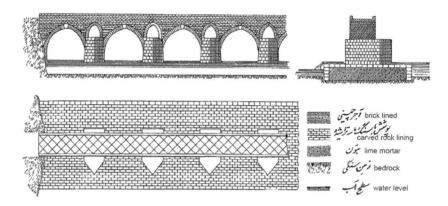

FIGURE 5.8   Schematic details of a Sasanian weir bridge.

5.9.2   POL-KHAJOO AND SIO-SE POL

Zayandeh Rud, the largest river in central Iran, traverses through the city of Esfahan, which was the capital of Iran during Safavid Empire. There are thirteen bridges over this river. The two oldest bridges are Pol-Khajoo (Khajoo Bridge) and Sio-se Pol (Thirty-three spans bridge).

Pol-Khajoo is the most appealing bridge in the Province of Esfahan. It was built by Shah Abbas II, a Safavian King around 1650, over the foundation of an older bridge. This bridge serves as both a bridge and a weir. It links the Khaju quarter of the city on the north bank with the Zoroastrian quarter to the south. Apart from its function as a bridge and a weir, it also served as a building, a place of public meetings. A pavilion exists at the center of the structure where Shah Abbas II would have once sat. Remnants of a stone seat is all that is left of this chair.

Pol-Khajoo is one of the finest examples of the Persian architecture at the height of the Safavian cultural era. Figure 5.9, a picture taken by the author in 1975, depicts this beautiful bridge which still is functional without a need for a major repair for three and a half centuries.

Sio-se Pol, also known as pol Allahverdi Khan, is the longest and the oldest bridge over Zayandeh Rud. It is nearly 298 m (977 ft) long and 13.75 m (45.1 ft) wide and consists of two superimposed rows of arches, 33 in the bottom, the longest of which is 5.60 m wide. It took three years from 1599 to 1602 to build this bridge. Its construction was supervised by Allah Verdi Khan, a Chancellor of Shah Abbas I. It is known that this man threw some silver coins in the wet lime mortar that formed the bridge foundation so that the workers keep compacting it by foot for search of coins. Figure 5.10, a photo taken by the author in 1975, shows this beautiful bridge, which after four centuries is still fully operational.

5.9.3   DAMS

Construction of dams in Iran dates back to Hakhamaneshian Empire, 2,500 years ago. One of the first dams which was built during Cyrus the Great (Kourosh Kabir or Bozorg) was located on the headwater of the Diyala River, in present Syria.

FIGURE 5.9   Pol-Khajoo over the Zayandeh Rud in Esfahan (Photo taken by the author).

FIGURE 5.10   Sio-se Pol (Allahverdi Khan) (photo taken by the author (1975)).

Three dams were built on the Kor River in Khuzestan by the order of Daruish Kabir. Of these, only one has survived. This dam is located downstream of the new Kourosh Kabir Dam, which was built during the last Pahlavi King (who reigned from 1941 to 1979). Figure 5.11 shows remnants of Daruish Kabir Dam, which is over 2,500 years old. This dam, as the picture indicates, was built using large blocks of rock and included low openings for the passage of normal flows.

FIGURE 5.11   Remnant of Daruish Kabir Dam on the Kor River (Adapted from Figure 140, Reza et al.).

The figure on the cover page of the book shows a gravity dam in Shiraz. This dam, named Bahman Dam (Sad-e Bahman) is well over 2,300 years old. It is known that this dam was built during Hakhamaneshian Empire and was repaired during the Sasanian Empire.

A multi-purpose dam, named Band Amir (Amir Dam), exits on the Kor River, nearly 35 km north of Shiraz in Fars Province. This beautiful dam, which includes a number of outlets and an emergency side channel, was built in the early part of the fourth century SH (tenth century AD) using granite blocks with lime mortar and was anchored with steel clamps cast in with lid. This dam is still operative and is one of the landmarks in Iran.

In addition to gravity and earth dams, Persians had also built arch dams. Band Kerit (Kerit Dam which is situated in Tabas, about 400 km southwest of Mashhad is one of the three ancient arch dams. This dam which was built over four centuries ago and is more than 100 m high, was built using bricks and now is filled with silt and clay.

5.10   AB PAKHSHAN (WATER DIVIDERS)

Some of the weirs and dams (which were named "bands" in Iran) were provided with openings to distribute water of a river among a number of irrigation canals. Such structures were called "ab pakhshan" meaning water dividers. To streamline the flow and distribute it uniformly, the divider piers, which were commonly made of stone, were constructed having a pointed triangular head. Alternatively, the dividers were raised to function as a weir to impound water behind the structure and uniformly distribute water in irrigation canals. Also, the bed of the stream was flattened at a moderate slope before the weir structure. One such water divider which was built on the Taraz River in Kashmar,[5] Khorasan; has several channels, each of which directs water to a village for irrigation needs.

5.11   CONSTRUCTION MATERIALS

In constructing irrigation structures such as dams, weirs, ab-pakhshans and canals, Persians used local materials to the maximum extent feasible. These materials included rock, clay mud, sun-dried clay bricks (khasht), baked clay bricks, lime mortar (sarooj), clay-lime mixture (shafteh), gypsum and metals.

5.11.1   CLAY MUD AND SUN-DRIED BRICKS

Clay muds were made mixing water with clay. Such muds, as was indicated by Karaji and indicated earlier in this chapter, were employed as an impermeable cover on the bottom of qanats, canals and irrigation channels. Where the clay mud was exposed to air, hay straw was added to the clay mud and mixed thoroughly. In this type of mud called "kahgal" (straw mud), the straw functioned as a reinforcement element and eliminated cracking and shrinkage. Such muds (reinforced earth) were impermeable once dried out and were used in constructing ab-anbars, ice houses and building walls. They were also used as a cover material on flat roofs. In such application a layer of "kahgal" 5–6 cm thick was placed uniformly on the roof area and compacted using a cylindrical hand roller made of stone. This type of cover lasted 25–30 years, functioned as an impermeable layer and served as a very good insulation material against summer heat and winter chill. They were used until the mid-twentieth century in Tehran and other cities and are still used in many rural homes in Iran. The roof area was run with a roller after heavy rains or upon removing snow to avoid swelling and to prolong the life of the roof.

Clay muds were also formed as square or rectangular sun-dried bricks and used for constructing building walls and covering canal beds. Until the 1930s, many homes in Tehran and large cities in Iran were constructed using sun-dried bricks and kahgel (clay mud mortar). Sunn dried bricks are still used in some villages. The home where I grew up, which was built in the 1910s, was constructed using sun-dried clay bricks and kahgel flat roof.

5.11.2   BAKED BRICKS

The inner surface of many ancient culverts, some qanats and outer surface of bridges, dams, ab-anbars and buildings were covered with baked clay bricks.

5.11.3   STONES

Both uncarved rocks and carved rocks were used in irrigation canals, dams and other hydraulic structures. In early stages rocks were used in their natural state to build walls, dams and to line canals. Later, clay muds were added as a mortar to paste the rocks together. In remnants of the buildings and dams, which were constructed twenty-five centuries ago during Achaemenid (Hakhamaneshian) Empire, one can see the use of carved and polished rocks. The tall cylindrical columns in Persepolis is a good example of the craftsmanship of carving and finishing large rocks. Figure 5.11 exemplifies the use of large unpolished carved and polished rocks for constructing weir bridges and water dividers (ab-pakhshans).

5.11.4   LIME MORTARS

Discovery of lime goes back to the Stone Era or perhaps even earlier. Lime was used in all ancient structures inclusive of water storage, irrigation and dwellings. To make this mortar the lime rock is heated in an oven to 600°C. At that temperature the rock loses its acid carbonic and becomes porous and 40% lighter. The porous rock is kept dry and away from absorbing any moisture. Then water is sprinkled on the porous lime rock to break it apart and is powdered with an iron tamper. The powder is kept thoroughly dry until it is used. Adding water to the lime powder forms a past of calcium hydrate. This paste can be shaped as desired and is used quickly before it hardens. Once exposed to air, the paste loses its water and turns back to its natural lime rock state. The more it is kept away from water and humidity and the quicker the mortar is applied when exposed to air, the sooner the mortar hardens. This type of lime is called air type lime (ahek havaband) because it hardens only when exposed to air.[6] Another type of lime is a wet lime mortar. Constructing foundations of bridges and dams, and piers, as well as lining the bed of canals and qanats, required a mortar that hardens in water. There is no record when Persians learned to make a wet mortar; however, the remnants of dams and bridges which were built twenty-five centuries ago reveal that Persians were already using such a mortar at that time. In European writings, the use of wet limes goes back to AD 400. It is evident that Persians, Creeks and Romans learned independently how to make a wet lime mortar and also their process of preparation differed (see the footnote for the mortar preparation in Rome).[7]

Karaji in his millennial book describes the process of making wet lime mortar which he refers to as "gray lime" (now called sarooj in Iran). His procedure of making this mortar was described earlier in this chapter. This mortar, which was used as late as the 1930s in Iran, was prepared adding ashes from bath houses or lime making furnaces and a small amount of fiber to lime powder. And, after mixing and pounding the paste thoroughly, it was used as an impermeable material in ab-anbars, bridge piers and small dams. The materials used in constructing Shushtar Hydraulic Structures were granite and sarooj. It is to be noted that today's cement, which was invented about 200 years ago, is in fact a modified type of the old wet lime. Of course, the new furnaces, which create temperatures as high as 1400°C, cause impurities in clay soil to turn into a molten paste. The remainder of clay soil and lime form solid balls the size of a hazelnut which is known as clinker in the cement industry. Upon cooling and milling, clinker cement is created.

5.11.5   CLAY-LIME MORTAR

This mixture is prepared mixing lime with clay and adding water to make a paste. During archeological excavation in Tapeh Hasanlu[8] from 1956 to 1974, a tomb was discovered dating back over 3,000 years ago. This tomb was 4 m deep and covered with a 3 m thick slab made of clay-lime mixture. Though there is insufficient information on its dimensions, the slab was large enough to cover the tomb which contained a dead person in sitting position and four horses. This finding is very important in that it indicates that Persians were using lime to reinforce clay mud thousands of years ago. The lime in this mixture functions the same way as that of the cement in concrete mix in that it bonds a much larger quantity of sand and gravel to make a strong construction material. The clay-lime paste called "shafteh" or "limey concrete" has been used in many ancient structures in Iran and used to be a common construction material until the invention of concrete. This material was not only strong, but also because of adding a small amount of lime which requires preparation to clay which is readily available, was very economical.

5.11.6 GYPSUM

The use of gypsum in Persia most likely dates back before lime mortar. The plaster rock is $SO_4Ca + 2H_2O$, which upon heating in furnace, loses its water and becomes lighter. When this rock is milled and water is added to the powder, it hardens quickly and forms gypsum. In the early 1960s, during a new excavation in Haft Tapeh (also written as Hafttape and Haft Tepe; literally means seven mounds) in Khuzestan, it was discovered that arch structures were common in Persia thousands of years ago and that gypsum mortar (plaster of Paris) was used for plastering walls and ceilings.[9] During the 1950s and 1960s, Haft Tapeh which is located approximately 10 km southwest of the ancient city of Susa (Shush, an ancient historic city in Khuzestan), became a site of large sugar cane plantation. (Replacing orchards by sugar canes has had an adverse impact in this area, author.) While leveling the land for planting, some archeological remains were destroyed and others exposed. An archaeological expedition began to explore the site in 1965 and found massive sun-dried and baked brick buildings.[10] The buildings in Haft Tapeh were constructed during a single period lasting one or at the most two centuries during the middle of the fourth millennium BC when Haft Tapeh was a major Elamite City during Bronze Age. For yet to be known reasons, the city began to decline, and some of the palaces and temples were abandoned. Later, some of their material was used to build dwellings.

The sun-dried bricks in the buildings there were pasted together using a clay mortar and the baked brick was pasted with extremely strong gypsum mortar. Gypsum was also used as a cover on baked brick pavement. An uncovered archeological remain revealed a large temple where the god Kirwashir was worshiped. Beneath the temple was a funerary complex intended for the King and his family and skeletal remains were found in the tomb.

Recent excavations, led by archaeologist Behzad Mofidi Nasrabadi of Mainz University in Germany, have uncovered a clay tablet archive. Also, behind a wall in this Bronze Age city remains of several hundred massacre victims were found (Archaeology, November 6, 2015).

In the royal tomb discovered in Haft Tepe, there was a very hard cover approximately 10 cm thick around which holes were made. This cover acted like a reinforced concrete ceiling. Figure 5.12 shows this tomb. Dr. Ezzatollah Negahban (1990), then a leading member of archaeological group, took a piece of this cover and sent it to Professor Dr. Ing. H. E. Schubert, the head of materials science and testing structural materials of the University of Munich in Germany, for testing. The tests, which were conducted on March 4, 1971, indicated that the cover was composed of 84% gypsum (SO_4Ca); 5% lime (CO_3Ca) and 4% soluble quartz, clay, iron oxide and magnesium.

FIGURE 5.12   Royal Tomb in Haft Tapeh.

In comparison to this 93% effective material, there was a 7% insoluble sand in the sample, implying that the plaster was very pure.

Samples were also taken from the delivered piece and tested for compression and flexural tensile strengths of the plaster piece. The measurements indicated the following (Reza et al., pp. 104–107):

- The measured compressive strengths of five samples varied from 64.2 to 98.5 kPa/cm^2 and indicated an average strength of 76.8 kPa/cm^2. In comparison, the standards at the time called for compressive strength of a plaster dried to 35° to 40°C to be equal to or higher than 60 kPa/cm^2. Therefore, the piece of ancient plaster at Haft Tapeh not only met but exceeded this standard. Dr. Schubert, based on the tests, noted that the plaster was superior to that used in Egypt's pyramids.
- The measured flexural tensile strengths of two samples were 27.2 and 30.8 kPa/cm^2. The measurements indicated an average strength of 29.0 kPa/cm^2. The structural standards in Germany DIN 1168 specified at a minimum 25 kPa/cm^2 of flexural tensile stress for the plaster mortar tested under the above indicated conditions. Thus, the plaster mortar which made several thousands of years ago in Persia adhered with the 1971 structural standards in Germany.

5.11.7 METALS

After the Stone Age which goes back to over 7,000 years ago, the metal age began. Copper, which is relatively soft and occurs as fairly pure as gold was the first metal used by humans. The next stage in metal use occurred during the Bronze Age, around 2,300 BC. Bronze is a mixture of copper and tin but harder than copper. The last stage in metal use is the Iron Age which goes back to approximately 700 BC. A difficulty in making iron was that it needed to heat up to 1400°C. to melt, and therefore, was difficult to make and was an expensive metal. Iron was used as anchors and fasteners in structures. In ancient hydraulic structures in Persia, iron was used

FIGURE 5.13   A column piece in Persepolis – note square hole and roughened center.

as anchors to fasten rocks together. The large stone blocks of columns in Persepolis were fastened using molten irons or lead in joints which consisted of square-shaped male and female ends. Figure 5.13, taken by the author in 1971, depicts a column segment with a square hole in Persepolis. The hole retains a male piece in the adjoining section, and the gap is filled with molten metal to hold the pieces tightly together. The roughened center, shown on the picture, increases friction and further helps avoid displacement of column pieces.

NOTES

1   This procedure is more practical in wide channels than narrow ditches and was commonly practiced in Iran before motorized rollers were made.
2   Eggs were also added to the mortar that was used for the footings of Sio-se Pol (Bridge) in Isfahan, nearly 400 years ago. This bridge is still standing.
3   Persians had at least one holiday every month. After Arabs invasion, these holidays were abandoned, only Norooz (incorrectly spelled Nowruz in the English language) is now observed by government and just a few other, such as Yalda night, are celebrated by people.

 4 SH = Solar Hejri.
 5 Kashmar is located approximately 260 km south of Mashhad, the Capitol of Khorasan, one of two provinces in Khorasan (renamed as Khorasan Rezavi by Islamic Republic).
 6 Karaji's procedure for making a version of lime mortar was described earlier in this chapter.
 7 Romans were using fired clay or brick dust or volcano ashes (called Puzzuolan) from the volcano in Puzzuoli coast at the foothills of Mount Vesuvius as an additive to lime powder. Mortars made with brick dust and lime were red colored and appear at many of the Roman buildings.
 8 Tapeh Hasanlu is located a short distance south of Lake Urmia now in the Province of West Azerbaijan. This site was an ancient center of civilization and was inhabited from the sixth millennium BC to the third century AD.
 9 Gypsum mortar had been used in the pyramids in Egypt as well.
10 The archaeological site covers approximately 1.5 km by 800 m.

Figure Sources

Figure 1.1 Wikimedia Commons (public figure).

Figure 3.1 Reza et al.

Figure 3.2 Reza et al.

Figure 3.3 Kaushik, https://www.amusingplanet.com/2015/02/the-wind-catchers-of-iran.html (Wikicommons).

Figures 4.1, 4.2, 4.3, 4.4, 4.8, 4.9, 4.13, and 4.14 Reza et al., 1971.

Figures 4.5, 4.6, 4.7, 4.10, 4.11, and 4.12 Karaji, 1966.

Figures 5.1, 5.4, 5.5, 5.6, 5.8, and 5.11 Reza et al., government publication.

Figure 5.2 Public domain/wikicommons.

Figure 5.7 S. Elemad, a relative.

Figures 5.9, 5.10, and 5.13 Photos by author.

Figure 5.12 Public domain/wikicommons.

References

Aber, J. S., Alberuni calculated the earth's circumference at a small town of Pind Dadan Khan, Emporia State University, District Jhelum, Punjab, Pakistan, Abu Rayhan al-Biruni, Emporia State University, 2003. https:/academic.emporia. edu/aberjame/histgeol/biruni/biruni.htm.

Agarwal, A., "Water and Sanitation for All?" Earthscan Press Briefing Document, No. 22, Earthscan, London, 1980, p. 53.

Amusing Planet, The Wind Catchers of Iran, by Kaushik, February 2015. https://www.amusingplanet.com/2015/02/the-wind- catchers-of-iran.html.

Ancient Origins, "Babak Khorramdin-The Freedom Fighter of Persia," https://www.ancient-origns.net/history-famous-people/babak-khorramdin-freedom-fighter-persia-002590/page/1/1.

Archaeology (A publication of the Archaeological Institute of America), "Massacre Victims Discovered at Iran's Haft Tapeh," Friday, November 6, 2015, https://www.archeology.org/ news/3862-151106-iran-haft-tappeh.

Asfazari, Abu Hatem Mozafar, "Surviving Treatise of Alavi," edited by Mohammad Taghi Modress Rezavi, Bonyad, Farhang, Iran, Publication No. 264, Khajeh Publishers, Tehran, 2536 Shahanshahi, 1977, p. 116.

Atlas Obscura, "Objects of Intrigue: Ancient Persian Water Clocks," by TaoTao Holmes, February 2, 2016, https://www.atlasobscura. com/articles/objects-of-intrigue-ancient-Persian-water-clocks.

Ayvan-e Kesra Enclopedia Iranica, https://www.iranicaonline.org/ articles/ayvan-e-kesra-palace-of-kosrow-at-ctesiphon.

Bahadori, M. N., "Passive Cooling System in Iranian Architecture," *Scientific American*, vol. 238, no 2 1978, pp. 144–154.

Baker, M. N., and R. E. Horton, "Historical Development of Ideas Regarding the Origin of Springs and Ground-Water," *Transactions American Geophysical Union*, vol. 17, 1936, pp. 395–400.

Bahadori, M. N., "Passive Cooling System in Iranian Architecture" *Scientific American*, Vol. 238 Feb. 1978, pp. 144–154.

Baker, M. N., and R. E. Horton, Historical development of ideas regarding the origin of springs and ground-water, *Trans. Amer. Geophysical Union*, Vol. 17, 1936, pp. 395–400.

Baker, S. W., *Cyprus as I Saw It in 1879* (book), London, Macmillan & Company, 1879, rare books, Available at Amazon and Barnes and Nobels.

Beadnell, H. J. L., "An Egyptian Oasis: An Account of the Oasis of Kharga in the Libyan Desert, with Special Reference to Its History," *Physical Geography and Water Supply*, 1909, p. 171.

Beadnell, H. J. L., "Remarks on the Prehistoric Geography and Underground Water of Kharga Oasis," *Geographic Journal*, vol. 81, 1933, pp. 128–131.

Beaumont, P., "Qanats on the Varamin Plain, Iran," *Transaction British Geographers*, Pub. No. 45, 1968, pp. 169–179.

Beaumont, P., "Qanat Systems in Iran," *Bulletin Intl. Assoc. of Scientific. Hydrology*, vol. 16, 1971, pp. 39–50.

Beckett, P., "Qanats Around Kirman," *Journal of Royal Central Asian Society*, vol. 40, 1953, pp. 47–58.

Behnia, A., Construction and Maintenance of Qanats, University Publication Center, Tehran, 1988, pp. 236.

Biruni, Abu Rayhan, *Surviving Chronologie*, translated in Farsi by Akbar Dana Seresht, Ibn Sina Publishers, Tehran, Iran, 1973, pp. 615.

Biswas, A., *History of Hydrology*, North-Holland Publication Co., Amsterdam, Holland, 1970, p. 336.

Bouwer, H., *Groundwater Hydrology*, McGraw-Hill, New York, 1978, pp. 480, 186–188.

Boyer, C. B. A History of Mathematics, Chapter 13, John Wiley and Sons, Inc, New York, NY, 1986

Boyle, A. J., *Cambridge History of Iran*, Cambridge University Press, Cambridge Vol. 5, 1968, p. 780.

British Admiralty, Persia, *Geographical Handbook, Series BR525*, London, 1945, p. 541.

Butler, M. A., "Irrigation in Persia by Kanats," *Civil Engineering Magazine, ASCE*, 3, 1933, pp. 69–73.

CAIS (The Circle of Ancient Iranian Studies), Ayvan (Taq)-e Khosrow, by E. J. Keall, http://www.cais-soas.com/CAIS/ Architecture/ayvan_e_Khosrow.htm

Clapp, F. G., "Tehran and the Elburz," *Geographical Review*, vol. 20 (1), 1930, pp. 69–85.

Cressey, G. B., "Qanats, Karez and Foggaras," *Geographical Review*, 48 (1), 1958, pp. 27–44.

Cyrus Charter of Human Rights-First charter of Human Rights, https://www.farsinet.com/cyrus/, in p. 7.

Cyrus the Great, by Cyrus Soral, https://cyrusthegreat.net.

Dampier, W. C., *A History of Science and Its Relation with Philosophy and Religion*, Cambridge University Press, Cambridge, Fourth Ed., 1949.

Davis, S. N., Discussion of "Exploration of Hidden Water" by Mohammad Karaji-The Oldest Textbook on Hydrology," *Groundwater Journal*, 11 (4), 1973, p. 45.

de Camp, L. S., *The Ancient Engineers*, Dorsel Press, New York, 1963. Encyclopediairnica Abyari "Irrigation in Iran," pp. 10 http://www.iranicaonline.org/articles/abyari-irrigation-in iran.

Encyclopedia Iranica, Haft Tepe, 6 pp. https://iranicaonline.org>articles

Encyclopedia Iranica, *Haft Tepe*, www.iranicaonline,org> hafttape, in 6 pages. 6.

Engheta, N., *Lion and Sun, Three Thousand Years Symbol*, Second Ed., In Farsi, Unlimited Publication Co., Los Angeles, CA, 2005.

English, P. W., "The Origin and Spread of Qanats in the Old World," *Proceedings of the American Philosophical Society*, vol.112, no.3 1968, pp. 171–181.

Estaji, H. and E Raith, "The Role of Qanat and Irrigation Networks in the Process of City Formation and Evolution in the Central Plateau of Iran, The Case of Sabzevar," Chapter 2 in *Urban Change in Iran*, Springer, New York, London, 2016, p. 425.

Farsi Net Cyrus Charter of Human Rights, Cyrus Cylinder–First Charter of Human Rights, http://www.farsinet.com/cyrus/, in 7 pp.

Fitt, R. L., "Irrigation Development in Central Persia," *Journal of Royal Central Asian Society, London, England,* vol. 40, 1953, pp. 124–133

Ghahraman, F., *The Right of Use and Economics of Irrigation Water in Iran*, University of Michigan, Ann Arbor, MI, 1958.

Giorgio Levi Della Vida, "Appuntie quesiti di storia tetferaria araba, 4. Due nuove del matematico al-Karagi (al-Karkhi)," *Revista degi Studi Orientali, Rome*, vol. 14, 1934, pp. 249–264.

Goblot, H., "Dans l'ancien Iran, Les Technîques de l'eau et la Grande Histoire," *Annales Economics Societes Civilization*, 18, 1963, p. 510.

Goldschmidt, A., *A Concise History of the Middle East*, Westview Press, Boulder, CO, 2002, pp. 76–77.

Goldsmith, E., "The Qanats of Iran," 2015, https://www.edwardgoldsmith.org/1031/the-qanats-iran/, p. 7.

Goldsmith, E., and N. Hildyard, *The Social and Environmental Effects of Large Dams*, Vol. 1, Wadesbridge Ecological Center, Worthyvale Manor Camelford, Cornwald, UK, 1984, Chapter 21.

Grewe, K., *Licht am Ende des Tunnels. Planning und Trassierung im Antiken Tunnelbau*, Mainz, 1998, pp. 94–96.

Haft Tappeh: The Elamite City of Kabank, *News Network Archaeology*, 10/31/2011 in 8 pages. https://archaeologyne-wsnetwork.blogspot.com/2011/10/haft-teppah-elamite-city-of-kabank.html

Helweg, O. J., "Discussion of "Exploration of Hidden Water," by Mohammad Karaji-The Oldest Textbook on Hydrology? *Groundwater Journal*, vol. 11, no.4, 1973, p. 44.

Humlum, J., "L' Agriculture par Irrigation en Afghanistan," *Comples Rendus Congre's International de Geographie*, vol. 3 (1951), 1949, pp. 318–328.

Karagi, Mohammad al, La Civilisation Des Eau Cachees Traite de l'Exploitation des eaux souterraines compose en 1017 AD, Texte ētablī, traduit et commente Par Aly Mazaheri, University of Nice Institute D'Etudes et de Recherches Interenthniques et Interculturelles (IDERIC), 31 Rue Verdi, 06000, Nice, 1973.

Karaji, Mohammad, Extraction of Hidden Waters (c. 1000 AD), translated to Farsi by H. Khadiv Jam, Bonyad Farband. Iran (Iranian Cultural Foundation) Publications, Tehran, Iran, 1966, p. 129 (Iranian Calendar 1345).

King, L. W., *The Code of Hammurabi*, translated to English, Yale University Press, New Haven, CT, 1915.

Kuros, Gh. R., *The Art of Irrigation and Dam Construction in Ancient Persia*, Iranian Cultural Foundation, Tehran, 1966, p. 127.

Lambton, Ann K. S., Landlord and Peasant in Persia, Oxford University Press, London, 1953, pp. 210–229.

Lambton, Ann K. S., The Persian Land Reform (1962-1966), Clarendon Press, Oxford, 1969.

Linsley, R. K. and Franz ini, J. B., *Water Resources Engineering*, McGraw-Hill, New York, 1972, pp. 104–112.

Linsley, R. K., M. A. Kohler and J. H. L. Paulhus, *Hydrology for Engineers*, Third Ed., McGraw-Hill, New York, 1982.

MacCurdy, E., *The Notebooks of Leonardo da Vinci*, Vol. 1, Reynal and Hitchcock, New York, 1939.

MacFarden, W. A., "The Early History of Water Supply; Discussion," *Geographic Journal vol.* 99, 1942, pp. 195–196.

Madelung, W., 1984, *Al-e Bavand (Bavandids), Encyclopedia Iranica*, Vol. 1, Fasc. 7, Routledge and Kegal Paul, London, pp. 747–753, ISBN90-DY-08114-3.

Marne, W. E., "The Ghanat, Ingenious Horizontal Well of Ancient Persia," *Desert Magazine of the Outdoor Southwest*, 1960, pp. 20–21.

Mays, L. W., Water Resources Handbook, McGraw-Hill, New York, 1996.

Meinzer, O. E., "The History and Development of Groundwater Hydrology," *Journal of the Washington Academy of Sciences*, vol. 24, 1934, pp. 6–32.

Merchel, C., *Ingenieur Technik in Altertum*, Spinger Verlag, Liepzig, 1889.

Merritt-Hawkes, O. A., *Persia, Romance and Reality, London,* Ivor Nicholos and Watson, London, 324 pp. book, 1935, pp. 115–117, 144–145, 152–153, 268, 295.

Nadji, M. and R. Voigt, "Exploration for Hidden Waters, by Mohammad Karaji, The Oldest Textbook on Hydrology?," *Groundwater Journal*, vol. 10, no. 5, 1972, pp. 43–46. (available at C.S.U.)

Narshakhi, Abobakr Mohammad-Ibn-Jafar, 1948, *History of Bokhara*, translated by Abo-Nasr Ahmad-Ibn-Nasr-Alghiavi, rewritten by Mohammad-Ibn-Zafar-Ibn-Omar, revision by Modares Razavi, Techran, Zoar.

Nasiri, F. and M. S. Mofakheri, "Qanats Water Supply Systems: A Revisit of Sustainability Perspectives," *Environmental Systems Research (Springer)*, vol. 4 no. 13, 2015.

Nasr, Seyed Husayn, *Islamic Scholars in Contemporary World* (in Farsi: Maaraf Islami in Jahan Mooaser), Pocket Books Corporation (Shraket Sahamie Books Jibi), Tehran, Iran. 1969, pp. 299.

Negahban, Ezatollah, *Excavation at Haft Tepe, Iran*, Univ. of Pennsylvania Press, Philadelphia, PA, 1990.

Noel, E. Colonel, "Qanats," *Journal of Royal Central Asian Society*, London, vol.31, 1944, pp. 191–202.

Payvand Iran News Photos: 5000-Year-Old Water System Discovered in Western Iran, June 4, 2014, p. 8, https://www.payvand.com/ news/14/jun/1024.html.

Pazwash, H., "Erosion of Land Surface, A Practical Review of the Subject," *1983 Symposium of Surface Mining Hydrology, Sedimentation and Reclamation*, Nov. 27–Dec. 2, 1983, University of Kentucky, Lexington, KY, pp. 189–194.

Pazwash, H., "Water Resources Utilization in Iran, in the Past," *Memoirs of the Faculty of Engineering University of Tehran*, Tehran no. 41), 1980, pp. 9–22.

Pazwash, H., *Role of Qanats in Reclamation of Kavirs (Deserts)*, Memoirs of the Faculty of Engineering, University of Tehran, Tehran, no. 44, May, 1982, pp. 1–11.

Pazwash, H., *Iran's Mode of Modernization, Greening the Deserts, Deserting the Greenery*, Civil Engineering, ASCE, March 1983, pp. 48–51.

Pazwash, H., *Translation of "Geohydrology" by Roger J. M. DeWiest*, John Wiley, 1965, Tehran University Press, Tehran, Publication No. 1495, 1975.

Pazwash, H. and G. Mavrigian, "A Historical Jewelpiece-Discovery of the Millennium Hydrological Works of Karaji," *Water Resources Bulletin*, vol.16, no.6, 1980, pp. 1094–1096.

Pazwash, H. and G. Mavrigian, "Millennial Celebration of Karaji Hydrology," *ASCE, Journal of Hydraulic Division*, 107 (H43), 1981, pp. 303–309.

Pazwash, H., *Urban Stormwater Management*, Second Ed., CRC Press, Boca Raton, FL, 2016, p. 684.

Provincial Map of Iran, Zoroastrian Heritage, by K. E. Eduljee, Map of Iranian Provinces, 2006, 2 pp., http://www.heritageinstitute. com/zoroastrianism/maps/IranProvinces.htm

Reza, E., G. Kurus, M. A. Amam Shushtari, and A. A. Entezami, *Water and Irrigation Technology in Ancient Iran*, The Ministry of Water and Electricity, Tehran, 1974, p. 299.

Sachau, E., "*Reise in Syrien und Mesopotamien*," Liepzig, 1883. Sarton, G., *A History of Science*, John Wiley, New York, 1980, pp. 535.

Schindler, A. H., "Beschrieburg Einiger Weing Bekanner Routen in Chorassan," *Z. Ges End*, Berlin, Germany, 1877.

Schram, W. D., "Literature on Qanats," http://www.romanaqueducts.info.

Singer, C., E. J. Holmyard and A. R. Hall (Editors), *A History of Technology*, Vol. 1, Oxford Univ. Press, 1954, pp. 532–534.

Singer, C., E. J. Holmyard, A. R. Hall and T. I. Williams (Editors), *A History of Technology*, Vol. II, Oxford Universty Press, London, 1956, pp. 666–667.

Smith, A., *Blind White Fish in Persia*, George Allen and Unwin Ltd., London, 1953, p. 231.

Smith, D. E., History of Mathematics, Dover Publications, Inc., New York, 1958, Vol. I, pp. 283–287; Vol. II, pp. 380–389, 446–448.

Tabari, Mohamad-Ibn-Jarir, *Tarikh-al-Rusul-va-Muluk (History of Prophets and Kings)*, translated by Abolghasem Payande, Institute of Iranian Culture, Tehran, 1973.

Tabari, The History of al-Tabari, *The Conquest of Iran AD 641-643/ AH 21-23 (illustrated ed.)*, Vol. 14, State University of New York Press, New York, 1994, p. xvii, ISBN 978U79142947.

Taton, K. Ancient and Medieval Science from Prehistory to AD 1450, in 4 volumes, translated by A.J. Pomerans, Basic Books Inc, New York, NY, 1963.

Tolman, C. F., *Groundwater*, First Ed., McGraw-Hill Book Company, 1937, p. 593, pp. 12–14.

Trapasso, L. M., "Water Wisdom of the Ancients," *ASCE Civil Engineering Magazine*, 1996, pp. 64–65.

UNESCO, "Aflaj Irrigation Systems of Oman," https://whc.unesco. org/en/list/1207/, accessed on October 25, 2009.

Woepcke, Franz, *Extraits du Fakhrí, traite d'algebre*, Paris, 1853, p. 152.

Wulff, H. E., "Qanats of Iran," *Scientific American*, 218 (4), 1968, pp. 94–105.

Wulff, H. E., *The Traditional Crafts of Persia*, The MIT Press, Cambridge, MA, 1966, pp. 249–254.

Zaghi, N. and Finnemore, E. J., "Discussion of "Exploration of Hidden Water-The Oldest Textbook on Hydrology," by Mohammed Karaji," Groundwater Journal, vol., 11, no. 4, 1973, p. 44.

Zarrinkoob, Abdolhossein, *Do Qarn Sokut, in Farsi*, Nineteenth Ed., Sokhan Publication, Tehran, 2005.

Zarrinkoob, Abdolhossein, *Two Centuries of Silence, translated from Farsi into English* by Paul Sprachman, Mazda Publishers, California, 2017, 316 pp., ISBN: 978-1-56859-260-2.

Index

For Product Safety Concerns and Information please contact our EU
representative GPSR@taylorandfrancis.com
Taylor & Francis Verlag GmbH, Kaufingerstraße 24, 80331 München, Germany

www.ingramcontent.com/pod-product-compliance
Ingram Content Group UK Ltd.
Pitfield, Milton Keynes, MK11 3LW, UK
UKHW021123180425
457613UK00005B/197